The Eurasian Beaver Handbook:
Ecology and Management of
Castor fiber

The Eurasian Beaver Handbook: Ecology and Management of *Castor fiber*

R. Campbell-Palmer, D. Gow, R. Campbell,
H. Dickinson, S. Girling, J. Gurnell, D. Halley,
S. Jones, S. Lisle, H. Parker, G. Schwab and
F. Rosell

CONSERVATION HANDBOOKS SERIES

Pelagic Publishing | www.pelagicpublishing.com

Published by Pelagic Publishing
www.pelagicpublishing.com
PO Box 725, Exeter EX1 9QU, UK

The Eurasian Beaver Handbook: Ecology and Management of *Castor fiber*

ISBN 978-1-78427-113-8 (Pbk)
ISBN 978-1-78427-114-5 (Hbk)
ISBN 978-1-78427-115-2 (ePub)
ISBN 978-1-78427-116-9 (Mobi)
ISBN 978-1-78427-117-6 (PDF)

This book should be cited as Campbell-Palmer, R. *et al.* (2016) The Eurasian Beaver Handbook: Ecology and Management of *Castor fiber*. Exeter: Pelagic Publishing, UK.

A CIP catalogue record for this book is available from the British Library.

Cover images: beaver, Graham Brown; landscape, Roisin Campbell-Palmer; culvert protection, Skip Lisle; piped dam, Gerhard Schwab.

Primarily funded by Colchester Zoo, the Welsh Wildlife Trust and the Wildlife Trust, with additional contributions from the Royal Zoological Society of Scotland, Bund Naturschutz, Telemark University College and Scottish Wildlife Trust.

Contents

Contributors

Róisín Campbell-Palmer, Field Operations Manager, Scottish Beaver Trial, and Conservation Projects Manager, Royal Zoological Society of Scotland, UK. PhD candidate, Telemark University College, Norway.

Dr Ruairidh Campbell, Wildlife Conservation Research Unit, Department of Zoology, University of Oxford, UK.

Helen Dickinson, Tayside Beaver Project Officer, Scottish Wildlife Trust, UK.

Dr Simon Girling, MRCVS, Head of Veterinary Services, Royal Zoological Society of Scotland, UK.

Derek Gow, Director, Derek Gow Consultancy Ltd, Upcott Grange Farm, Lifton, Devon, UK.

Professor John Gurnell, Queen Mary University of London, UK.

Dr Duncan Halley, Norwegian Institute for Nature Research, Norway.

Simon Jones, Project Manager, Scottish Beaver Trial, and Director of Conservation, Scottish Wildlife Trust, UK.

Skip Lisle, President, Beaver Deceivers International, USA.

Professor Howard Parker, Telemark University College, Norway.

Professor Frank Rosell, Telemark University College, Norway.

Gerhard Schwab, Bibermanagement Südbayern, Wildbiologe, Germany.

Foreword

I remember my first real contact with beavers in the wild, 35 years ago, when I was staying on a farm in Jamtland in Sweden. My host, Erik, was a hunter and a farmer who also worked in the local town; his family farm was in beautiful countryside with ospreys, goldeneye ducks and cranes breeding in the bogs; in the forests lived elk and beavers. I remember thinking it was as Scotland should be. One evening, I walked to the sluggish river which ran nearby – and, after a mosquito-tormented stalk, I saw my first beaver. It sensed me, though, and with a slap of its tail it was gone. I sat down on a jumble of beaver-felled birch trees, but my wait was in vain. I asked Erik what he thought about the birch trees felled by the beavers across his track. His reply was so sensible: 'I wait until winter and then drive down with my tractor and trailer and log up those trees – they are nicely seasoned and ready for my log store. And sometimes I hunt one; would you like beaver for supper tomorrow? I'll get some out of the deep freeze.' I thought it tasted good, nicely braised – something between brown hare and roe.

I liked the matter-of-fact way in which he lived with the beavers but he also recognised their value in the wetland ecosystem. That value is what this excellent book is about; it's written by 12 experts who have brought together a wealth of experience and, most importantly, a mine of information on how we can learn to live with beavers again in the United Kingdom.

I'm so pleased that beavers are back in our country, because I recognise they are essential in helping to manage natural wetland ecosystems. It's been a long time coming, and much longer than I expected when I was part of the first serious discussions as a main board member of Scottish Natural Heritage back in the early 1990s. A successful scientific trial has been carried out by the Royal Zoological Society of Scotland and Scottish Wildlife Trust on land owned by Forestry Commission Scotland, and with scientific input from Scottish Natural Heritage. Beavers, being beavers, have also cropped up on their own in the Tay catchment of Scotland and occasionally elsewhere in Britain. The Scottish Government will soon, I hope, make the decision that this important keystone species should once again be part of our natural fauna, and we can see them restored to their original haunts.

During my wanderings looking at beavers and talking with beaver experts in a variety of European countries, I have been impressed by their knowledge and their understanding of the species within their home countries. They take a common-sense approach of working and living with an animal that can spring a few surprises.

In the pages that follow, the reader can get everything they need to know about the history and ecology of beavers, their impacts on human operations and their value within the ecosystem. This restoring of functioning ecosystems in such a fragile world will be more and more at the centre of our nature-conservation ethos. The authors write on many aspects of how to manage beavers and our activities; from proactive dam management to novel methods of reducing their impact on our interests. There is information on trapping, translocation, culling and many other subjects. It's always been my view that, to have beavers back throughout Britain, we will need robust management and a rapid response to those who ask for help to act against beavers that cause problems.

The main part of this handbook finishes with learning to live with beavers – and that could be the most important aspect. I believe that it's crucial for farmer, fisher, forester or any of us to learn to accept that although some species may at times cause us problems, we should remember the myriad species and actions of the natural world that allow us to farm and fish, to grow crops and trees, and to have fresh water and breathe fresh air. We, like the beavers, are part of the great ecosystem we call the Earth. Let us celebrate the return of the water engineer supreme.

Roy Dennis, 2015
Highland Foundation for Wildlife

Rhagair

Rydw i'n cofio fy nghyswllt cyntaf erioed gydag afancod yn y gwyllt, 35 mlynedd yn ôl pan oeddwn i'n aros ar fferm yn Jamtland yn Sweden. Roeddwn i'n aros gydag Erik, heliwr a ffermwr a oedd hefyd yn gweithio yn y dref leol. Roedd ei fferm deuluol yng nghanol cefn gwlad bendigedig gyda gweilch y pysgod, hwyaid llygaid aur a garanod yn magu ar y corsydd ac, yn y coedwigoedd, roedd elcod ac afancod yn byw. Rydw i'n cofio meddwl mai fel hyn y dylai'r Alban fod. Un noson, fe gerddais i at yr afon a oedd yn llifo'n araf heibio ac, ar ôl cael fy erlid gan fosgito, fe welais i fy afanc cyntaf; ond roedd wedi synhwyro fy mod i'n nesáu ac ar drawiad ei gynffon, roedd wedi diflannu. Fe eisteddais i i lawr ar bentwr o goed bedw wedi'u torri gan yr afanc, ond ofer fu'r aros. Holais Erik am ei farn am y coed bedw wedi'u torri gan yr afancod ar draws ei lwybr. Roedd ei ymateb mor ddoeth: 'Rydw i'n aros tan y gaeaf ac wedyn yn gyrru i lawr gyda 'nhractor a'r trelar ac yn casglu'r coed – maen nhw'n barod yn hwylus iawn i mi ar gyfer fy stôr o goed. Ac weithiau fe fydda' i'n hela un. Fyddet ti'n hoffi cael afanc i swper nos fory? Mi dynna i un allan o'r rhewgell.' Roeddwn i'n meddwl ei fod yn neis, wedi'i goginio'n dda, rhywbeth rhwng ysgyfarnog ac iwrch.

Roeddwn i'n hoffi'r ffordd ddi-lol yr oedd Erik yn byw gyda'r afancod ac roedd hefyd yn cydnabod eu gwerth i'r ecosystem tir gwlyb. Am y gwerth hwnnw y mae'r llyfr rhagorol yma'n sôn; mae wedi'i ysgrifennu gan 12 o arbenigwyr sydd wedi dod â chyfoeth o brofiad at ei gilydd ac, yn bwysicach, stôr o wybodaeth am sut gallwn ni ddysgu byw gydag afancod unwaith eto ym Mhrydain Fawr.

Rydw i mor falch bod afancod yn ôl yn ein gwlad ni oherwydd rydw i'n sylweddoli eu bod nhw'n hanfodol i helpu i reoli'r ecosystemau naturiol o dir gwlyb. Mae wedi cymryd amser maith – llawer mwy nag yr oeddwn i wedi'i ddisgwyl pan gymerais i ran yn y drafodaeth ddifrifol gyntaf fel aelod o fwrdd Scottish Natural Heritage nôl ar ddechrau'r 1990au. Mae arbrawf gwyddonol llwyddiannus wedi cael ei gynnal gan Gymdeithas Sŵolegol Frenhinol Caeredin ac Ymddiriedolaeth Natur yr Alban ar dir sy'n eiddo i Gomisiwn Coedwigaeth yr Alban, a gyda chyfraniad gwyddonol gan Scottish Natural Heritage. Mae afancod, oherwydd eu natur, wedi dod i'r golwg ar eu liwt eu hunain yn nalgylch Tay yn yr Alban hefyd, ac mewn mannau eraill ym Mhrydain yn achlysurol. Y gobaith ydi y bydd y llywodraeth yn gwneud penderfyniad yn fuan ac yn datgan y dylai'r rhywogaeth bwysig hon fod yn rhan unwaith eto o'n ffawna naturiol ni, ac y gallwn weld ei hadfer unwaith eto yn ei chynefin gwreiddiol.

Wrth i mi grwydro i astudio afancod a siarad gydag arbenigwyr ar afancod mewn gwledydd Ewropeaidd amrywiol, rydw i wedi synnu at eu gwybodaeth am y rhywogaeth a'u dealltwriaeth ohoni yn eu gwledydd. Maent yn defnyddio synnwyr cyffredin ac yn gweithio ac yn byw gydag anifail sy'n gallu synnu dyn ar adegau.

Mae tua chant o dudalennau i'w darllen yma ac fe gewch chi ddysgu popeth y mae arnoch angen ei wybod am hanes ac ecoleg afancod, eu heffaith ar weithrediadau dyn a'u gwerth yn yr ecosystem. Bydd adfer ecosystemau mewn byd mor fregus yn dod yn rhan fwyfwy canolog o'n hethos ni ym maes cadwraeth natur. Mae'r awduron yn ysgrifennu am sawl agwedd ar sut mae rheoli afancod a'u gweithgareddau; o reoli argaeau'n rhagweithiol i ddulliau newydd o leihau eu heffaith ar ein buddiannau ni. Ceir gwybodaeth am ddal, trawsleoli, difa a phynciau niferus eraill. Rydw i wedi bod o'r farn erioed bod rhaid wrth gyfundrefn reoli gadarn os am ddod ag afancod yn ôl ledled Prydain Fawr, ac os am ymateb yn gyflym i'r rhai sy'n gofyn am help wrth i afancod greu problemau iddynt.

Mae prif ran y llyfryn hwn yn cloi gyda dysgu byw gydag afancod, a hon fydd yr agwedd bwysicaf o bosib. Rydw i'n credu ei bod yn hanfodol i ffermwyr, coedwigwyr, pysgotwyr a phob un ohonom ni dderbyn y bydd rhai rhywogaethau'n achosi problemau i ni ar adegau, ond bod rhaid i ni gofio bod angen llu o rywogaethau a gweithredoedd ym myd natur er mwyn galluogi i ni ffermio a physgota, tyfu cnydau a choed, a chael dŵr ffres ac awyr iach i'w anadlu. Rydyn ni, fel yr afancod, yn rhan o'r ecosystem fawr yr ydyn ni'n ei galw'n Ddaear. Mae'n rhaid i ni ddathlu dychweliad y meistr ar beirianneg dŵr.

Roy Dennis, 2015
Highland Foundation for Wildlife

Acknowledgements

The authors would like to thank all those who have contributed to this collaboration. The information included here is a culmination of many years' worth of field work, animal management and landowner experience. We are grateful for the additions from Thomas Borup Svendsen (Danish Ministry of the Environment, Denmark), Mike Callahan (Beaver Solutions LLC), Bryony Cole (University of Exeter), Sean Dugan (Scottish Fisheries Co-ordination Centre), Gidona Goodman (University of Edinburgh), Stuart Jenkins (RZSS), Robert Needham (University of Southampton), Romain Pizzi (Zoological Medicine), James Scott (SNH), Janne Sundell (University of Helsinki), Julia Coats (APHA), Kelsey Wilson, Ian McGowan (SNH), Colin Seddon (SSPCA), Glenn Iason (James Hutton Institution), Stefanie Venske (NaturErlebnisZentrum Wappenschmiede), Jika Uhlikova (Nature Conservation Agency of the Czech Republic), Ales Vorel (Czech University of Life Science) and Charlie Wilson. The images and photographs included have been vital to illustrate various aspects of this manual, so many thanks are given to all those who contributed, particularly Rachael Campbell-Palmer and Karlene Hill (SWT). Lastly, many thanks are given to those who edited earlier drafts of this manual, their contributions have been invaluable. Special thanks go to Nick Warren, Martin Gaywood (SNH) and Ivor Normand.

1. Introduction

Beavers (i.e. the Eurasian beaver, *Castor fiber*, and the North American – sometimes referred to as Canadian – beaver, *C. canadensis*) are unique mammals that often capture people's fascination. These are large rodents, with specialised features such as their flat scaly tails, and distinctive behaviours including tree-felling and dam-building. Few other animals, apart from humans, have the ability to modify so drastically their surrounding environments. Beavers play a key role in wetland ecology and species biodiversity, providing vital ecosystem services including habitat creation, water management and quality improvement, and sediment retention. At the same time, these beaver activities can also present real challenges for land and wildlife managers.

The history of the Eurasian beaver represents an important case study for conservation and reintroduction programmes. By the late 19th century, the once widespread Eurasian beaver was reduced to a handful of relict populations in fragmented refugia across Europe, in which potentially 1,000–2,000 individuals survived (Nolet and Rosell 1998). As beaver populations dwindled, so did the understanding of living with this species pass from common knowledge. In the beavers' absence, landscapes continued to be altered by humans, and riparian environments in particular were engineered to suit agricultural and industrial needs. By the early 20th century, naturalists tended erroneously to believe

Figure 1.1 Eurasian beaver feeding on water lilies at the Scottish Beaver Trial. (P. Price)

that beavers were solely a species found in 'open woods alongside rivers, old river beds and lakes' (van den Brink 1967) and limited in distribution by habitat suitability.

Since then, Eurasian beavers have been restored to much of their former native range through proactive reintroductions and translocations (Halley *et al.* 2012). Contemporary experience of expanding beaver populations across Europe and North America demonstrates clearly that beavers can readily modify even heavily engineered landscapes to suit their own ecological requirements. The two extant beaver species, the Eurasian and the North American, inhabit wetlands and water bodies from north of the Arctic Circle, where they can endure five months of darkness and ice, to the everglades of subtropical Florida. The dry, arid environment of the Ulungaur watershed in Mongolia contains one of the last remaining Far Eastern beaver populations. Intensively utilised, cultural landscapes dominated by agricultural production with amenity woodlands, recreational areas and engineered water bodies are relatively unchallenging environments for beavers. Although it has been well demonstrated that environmental factors such as topography, hydrology and vegetation influence beaver distribution (Schwab *et al.* 1992; Rosenau 2003; Rosell *et al.* 2005), they are clearly a much more adaptive species than was initially believed.

The Eurasian beaver is a well-studied species capable of providing biodiversity and economic benefits through its natural activities. Its restoration is considered internationally to be a clear conservation success (Halley and Rosell 2002). While initially some countries (such as Finland and Russia) restored beavers to support a commercial fur trade, the majority of recent reintroductions have been implemented for nature conservation purposes. This emphasis has been prompted by a greater awareness of the ecological benefits which accrue from the presence of beavers.

The return of the beaver through a combination of reintroductions and natural recolonisation has often been viewed as a novel phenomenon. When beaver populations initially re-establish, the physical impact of their activities is often confined to a small

Figure 1.2 Beaver-occupied site (>20 years), Norway. The beaver lodge is located on the small islet near the centre of the photograph. The lodge is 2 km downstream from a hydroelectric dam so water levels can vary abruptly. Although beaver signs are evident on the ground, the overall tree structure remains largely unaffected, and many people are unaware of the beavers' presence. (D. Halley)

Figure 1.3 Tree-coppicing near the Deutsches Museum, in downtown Munich, an area first occupied by beavers around 2000. (R. Campbell-Palmer)

Figure 1.4 Signs of beaver activity along an urban canal (Freising, Bavaria) where house-owners regularly feed beavers. (R. Campbell-Palmer)

group of land-users such as farmers, foresters or water authorities (Siemer *et al.* 2013). As beaver populations increase, the novelty of their presence can be replaced by hostility from wider elements of society when more visible impacts occur such as the felling of specimen trees in public parks, orchards or gardens. It is inevitable that a process whereby people 'rediscover' what it means to live with beavers will become a critical component of coexistence. Understanding, tolerance and a willingness to manage undesirable aspects of beaver activity competently will also be of fundamental importance.

Figure 1.5 Beaver-created wetland habitat in agricultural areas in Bavaria. Various beaver-management techniques may be required, especially in modified environments. These may include flow devices, fencing, land purchase for conservation, trapping and removal. Where these can be planned and implemented, beavers can diversify an area and provide biodiversity hotspots in close proximity to other human land uses. (R. Campbell-Palmer)

1.1   Aims and purpose of this handbook

This handbook considers a broad range of issues which are likely to arise over time as beavers are re-established, particularly with respect to British landscapes, but also elsewhere in Europe. It affords a practical overview of the implications of beaver restoration and the management requirements. Most of the information presented here addresses the experiences gained from beaver restoration in Europe, but draws from practical experiences in North America where beaver populations have also recovered. The handbook describes the animals' field signs, lifestyle, their effects on the environment and appropriate mitigation techniques, as supposed to any wider-scale, long-term management strategy at a national level. In relation to Britain, the Eurasian beaver has not yet been formally reintroduced despite being an Annex IV species on the European Habitats Directive, although licensed trial reintroductions have occurred in Scotland (Scottish Beaver Trial) and most recently in England (Devon Beaver Trial). If the decision is made that beavers should remain, then it is anticipated their domestic legal status will change, and management strategies will be developed. We recommend that advice and necessary permission should be sought from the appropriate Statutory Nature Conservation Organisations (SNCOs) before employment of any mitigation measures.

Key concepts
- Beavers modify their environment and play a key role in wetland ecology and biodiversity.
- Beavers have been restored to much of their former native range from near-extinction following human exploitation.
- Understanding of beaver ecology, tolerance and a willingness to competently manage undesirable aspects of their behaviours are fundamental to living with this species again.
- This handbook discusses the implications of beaver restoration and likely management requirements in a British context.

2. The history of beavers in Britain

'The church … the mill, bridge, salmon leap, an orchard with a delightful garden, all stand together on a small plot of ground. The Teivi has another singular particularity, being the only river in Wales, or even in England, which has beavers. In Scotland they are said to be found in one river, but are very scarce.' Giraldus Cambrensis (1180)

The above quotation describes a British landscape, moulded by people to suit their needs, which was also occupied by Eurasian beavers. The salmon (*Salmo salar*) leap, where fish could be harvested for consumption during their annual migration, offers a tantalising clue as to why the Eurasian beaver may have survived on the River Teivi at a time when it had otherwise become such an unfamiliar species in a British landscape. In 10th-century Wales, a beaver pelt was worth 120d (pence), and a pine marten (*Martes martes*) pelt 24d, while Eurasian otter (*Lutra lutra*), Eurasian wolf (*Canis lupus lupus*) and red fox (*Vulpes vulpes*) pelts were worth 8d each (Rodgers 1947–48). Did this community, which successfully managed its fish, also tolerate beavers for the prospect of a similar and profitable annual harvest?

The Eurasian beaver is a native species which was once widespread in freshwater habitats throughout Britain. Fossil records indicate that beavers were present in Britain 2 million years ago – and archaeological evidence, including gnawed timber, dams, lodges, burrows and bones, has been recorded from a number of sites (Macdonald *et al.* 1995; Coles 2006). Various place names, illustrations and other references bear testament to the former presence of this animal (Coles 2006). In 1837, the nearly complete skeleton of a beaver was found in a hole in the bank of the River Stour, near Keynston Mill in Dorset.

Eurasian beavers were hunted to extinction in Britain for their castoreum (a secretion from their castor glands with supposed medicinal properties, used by beavers in chemical communication), meat, fur and other body parts such as jaw bones. By the 15th century, the trade in their fur was no longer economically viable due to overexploitation, and their presence began to fade from folk memory. Oral traditions of their presence on Loch Ness and in Lochaber in Scotland, and in the Ogwen Valley in Wales, lingered on until the end of the 18th century (Coles 2006). Some of the latest physical evidence of beaver presence in Britain is a beaver-chewed stick carbon dated to 1269–1396, which was found in the upper Tyne catchment (Manning *et al.* 2014). While we will never know when the last British beaver was killed, a church ledger record from the village of Bolton Percy near York in 1780 records the payment of a bounty of tuppence for the head of a beaver (Coles 2006). No other references to the species' survival beyond this date are currently known.

Figure 2.1 Beaver-related place names exist throughout Britain. (D. Gow)

2.1 Beaver reintroduction in Britain

The concept of reintroducing beavers to mainland Britain (they are not known to have ever been native to Ireland or any of the outer isles) is not new. Their restoration has been widely discussed in both the general media and academic literature. Despite recommendations for a more comprehensive process of restoration (Gurnell *et al.* 2009), the return of the beaver has been a haphazard affair. At the time of writing, two officially licensed trial beaver releases into the wild exist: the Scottish Beaver Trial in Mid-Argyll, and the Devon Beaver Trial on the River Otter, Devon. The Scottish Government is due to consider the results of the trial and make a final decision on the future of beavers in Scotland (expected some time in 2016). Additionally, in the mid-2000s it became apparent that a sizeable population of unlicensed wild beavers were distributed throughout the catchment of the Rivers Tay and Earn in Perthshire, Scotland (Campbell *et al.* 2012a). In England and Wales, there is scattered evidence from field signs and photographs of the existence of small numbers of free-living wild beavers. Wild beavers are therefore already present in Britain. There are also a growing number of enclosed populations, and there are proposals for further trial releases in England and Wales (e.g. Welsh Beaver Project). Unless British governments decide to remove these beavers their populations will increase and their distribution expand. An understanding of their management requirements will therefore become a necessity.

Lord Onslow, writing in the *Countryman* magazine in 1939, suggested that as the 'beaver have become extinct in England only within the last few centuries ... there seems no reason at all why they should not be reintroduced' (Onslow 1939). In 1994, Scottish Natural Heritage (SNH), Natural England (then English Nature) and Natural Resources Wales (then Countryside Council for Wales) began to consider the potential for restoring beavers in Britain. To date, this process has progressed furthest in Scotland, with an official trial reintroduction and the scientific study of the larger unlicensed population in Tayside, though the Devon Beaver Project was given a licence for a five-year scientifically monitored trial release on the River Otter in 2015.

In 1998, SNH launched a public consultation to ascertain wider attitudes to beaver reintroduction in Scotland (SNH 1998). While the majority of respondents favoured reintroduction, there were concerns and opposition expressed by others who feared that beaver activity would have detrimental impacts on farming, fisheries and forestry.

Figure 2.2 Beaver pair released as part of the Scottish Beaver Trial. (Scottish Beaver)

In response to these findings, SNH took the decision to develop a time-limited, trial reintroduction in the Forestry Commission Scotland (FCS)-managed Knapdale Forest in Mid-Argyll, to explore the feasibility of releasing beavers and to study their impacts. Although an initial SNH licence application (2002) was turned down in 2005 by the then Scottish Executive, a joint licence application submitted in 2007 by the Scottish Wildlife Trust (SWT) and the Royal Zoological Society of Scotland (RZSS) was successful. This led to the development of the Scottish Beaver Trial, the first licensed and unfenced reintroduction of Eurasian beavers (and in addition the first official mammal reintroduction) in Britain.

Figure 2.3 Camera-trap image of wild beaver on the River Otter, England. (T. Buckley)

Although to date the reintroduction process has progressed furthest in Scotland, feasibility studies on beaver reintroduction to Wales (Halley *et al.* 2009) and England (Gurnell *et al.* 2009) have been published. Since the first release at Ham Fen in Kent in 2002 (Campbell and Tattersall 2003), breeding populations of beavers have been established in large enclosures at several locations in England, Scotland and Wales. One example of a well-studied project of this type is the demonstration site established in 2011 in the upper watershed of the River Wolf which has been developed by the Devon Wildlife Trust (DWT 2013). These controlled projects can provide useful information when employed specifically to assess the role of beavers as habitat-modifiers and as tools for landscape management or stakeholder engagement.

Key concepts

- The Eurasian beaver is a native mammal to Britain, and was once widespread.
- Beavers were hunted to extinction for their meat and fur, and were largely extirpated by the 16th century.
- Beaver reintroduction has a long history in Britain. Currently, there are two licensed trial releases (Mid-Argyll and Devon), with free-living beavers present in Perthshire and parts of England.

3. Beaver biology and ecology

3.1 Taxonomy and distribution

The Eurasian beaver and the North American beaver are the only surviving members of the once larger family of Castoridae. Both modern beaver species are physically very similar, making them hard to distinguish in the field. They have very similar ecological requirements and behavioural patterns, and were once considered to be a single species. Chromosome analysis has identified that the Eurasian beaver possesses 48 pairs of chromosomes, while the North American beaver has 40 (Lavrov and Orlov 1973). As a result, the two species will not interbreed and produce viable offspring, even when attempted through captive breeding. Through differences in tail shape and subtle differences in their pelage, beaver species can be determined on closer inspection. Examination of the anal gland secretions provides reliable differences in colour and viscosity which can be used to determine beaver species and sex (Rosell and Sun 1999). It is now believed that divergence occurred about 7.5 million years ago when beavers first colonised North America from Eurasia across the land bridge of the Bering Strait (Horn et al. 2011). From the Eurasian fossil record of the Early and Middle Pleistocene (~2.4–0.13 Ma ago), in mature rivers and wetlands, modern beavers appear to have lived alongside, or possibly to have been locally excluded by, the slightly larger extinct beaver *Trogontherium cuvieri*; the prevalence of the two forms at archaeological sites demonstrates an inverse relationship (Mayhew 1978).

By the 16th century, Eurasian beavers were largely extinct across most of Europe and Asia. At its lowest population point, it is believed that the Eurasian beaver in its western range was reduced to 200 animals on the River Elbe in Germany, 30 on the River Rhone in France and ~100 in Telemark, Norway (Nolet and Rosell 1998). Since the 1900s, beaver numbers have recovered throughout much of their former European range as a result of a combination of legal protection, hunting regulation, proactive reintroductions/ translocations and natural recolonisations. Breeding farms that produced beavers for commercial fur farming, and later for release into the wild for restocking, were established at Voronezh (Russia) and Popielno (Poland) in 1933 and 1958, respectively (Jaczewski et al. 1995). The first official conservation translocation occurred from Norway to Sweden in 1922, and since then there have been more than 200 recorded translocations of beavers (Halley et al. 2012). Eurasian beavers have now been restored to over 24 countries in Europe (Halley and Rosell 2002) and are currently estimated to number over 1 million individuals globally (Halley et al. 2012).

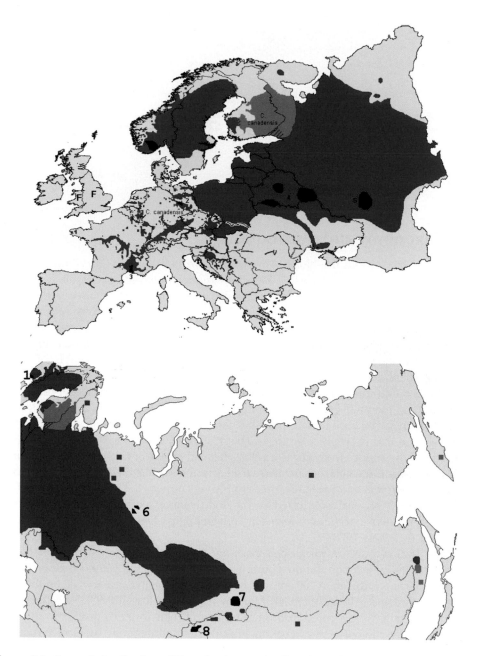

Figure 3.1 Current distribution of Eurasian beaver (red) and North American beaver (green). Black represents refugia where Eurasian beavers were never extinct (numbered west to east); 'F' indicates countries where reintroduction feasibility studies have been conducted. (Updated from Halley, Rosell and Saveljev 2012)

Also notable is a sizeable population (~8,000) of introduced North American beavers in Finland (Kauhala and Turkia 2013) and northwest Russia. In 1937, seven North American beavers from New York State, USA, were introduced to Finland to supplement an ongoing reintroduction of the nearly extinct Eurasian beaver there (Lahte and Helminen 1974). At that time, most zoologists recognised only one species of beaver worldwide (Parker *et al.* 2012). North American beavers are now also present in small numbers in parts of Belgium, western Germany and Luxembourg, as a result of escapes from a zoo and game parks (Dewas *et al.* 2012). Due to the significant biological and ecological similarities of these two species, identification and removal of the non-native North American beaver is important, requiring active management and resource investment (Parker *et al.* 2012).

3.2 Anatomy and appearance

Eurasian beavers are large (adults >20 kg), semi-aquatic, herbivorous rodents with a head and body length of ~80–100 cm and a tail length of ~30 cm on average in adults (Żurowski and Kasperczyk 1988), making them the second largest rodent species in the world. Beavers do not attain adult size until around three years of age. Adults of both sexes have similar head and body lengths, although females are on average 1–1.5 kg heavier than males, and are impossible to distinguish visually unless a female is heavily pregnant or lactating, during which time nipples will be visible (four under the forelegs).

Figure 3.2 Adult Eurasian beaver. (I. Sargent)

Coat colour is usually brown but can range from pale golden to black. The beaver's flattened tail is probably its most distinctive feature, covered in visible scales and dark grey in colour. When moving on land, beavers normally walk on all fours, although they can walk upright on their hind legs for very short distances when carrying mud or vegetation, particularly when building or restoring lodges for winter. During feeding sessions, they can be observed resting on their back legs while holding food in their forepaws. Beavers have small ears but have good hearing. Their eyesight is poor and largely functions to identify and respond to movements. They have an excellent sense of smell, and their chemical communication abilities are highly developed (Campbell-Palmer and Rosell 2010). While swimming, they utilise their powerful, webbed hind feet in a coordinated kicking motion for propulsion through the water and their tail as a steering rudder (Wilsson 1971; Novak 1987).

3.3 Breeding and young

Beavers live in family groups, usually comprising an adult breeding pair with their offspring from the current and the previous year's litters. Once paired, beavers tend to remain together until one partner either dies or is displaced by another individual of the same sex in territorial disputes. Mating usually takes place late December to February (depending on location). After a gestation of 105–107 days, the female gives birth to one litter per year, typically of one to four kits, in spring/early summer (with some geographical variation). The number of kits produced, and their survival, is affected by various factors such as the age of parents, the surrounding population density and habitat quality and altitude (Novak 1987; Campbell 2010; Campbell *et al.* 2013). Beaver kits are born fully furred, and usually weigh between 300 and 700 g. Their eyes are open soon after birth; and, although they will feed on their mother's milk for 2–3 months, they can consume vegetation after their first week. The kits remain in the lodge for approximately the first 1–2 months of their lives while their parents and older siblings bring leafy twigs and other vegetation for them to eat. By the time they are approximately 2 years old, they have become sexually mature and usually disperse in order to attempt to establish territories and partners of their own (Wilsson 1971).

Although some beavers remain with their family units as non-breeding individuals for many years, particularly if there is a lack of suitable habitat to move to (Campbell *et al.* 2005), most 2 year olds will begin to search for territories of their own in the spring. During this period, they are capable of travelling long distances along water bodies and may undertake shorter trips overland. Beavers do not like to be far from the water or to

Figure 3.3 Mother and ~3-month-old kit. (S. Gardner)

travel over open land for large distances, though reports of individuals up to 11.7 km away from water exist these are rare (Saveljev *et al.* 2002); the barrier effect of watershed divides on beaver colonisation varies depending on topography (Halley *et al.* 2012). There is no consistent difference in dispersal between the sexes (Saveljev *et al.* 2002; McNew and Woolf 2005), and individuals will commonly make exploratory excursions into neighbouring areas (Campbell *et al.* 2005). After dispersal, they have been recorded taking up residence in some unconventional aquatic habitats, such as small ornamental ponds or sewage-treatment plants. The dispersal process can be hazardous for beavers, with deaths via road traffic accidents or through infections from wounds caused by other beavers. Beavers from one family are highly intolerant of those from other families and will fight aggressively with intruders. During these fights, deep, penetrating wounds are frequently inflicted by biting to the shoulders, flank and tail (Figure 3.4). Such wounds can result in severe injuries and/or become septic, and can result in death (Piechocki 1977).

Figure 3.4 Crescent-shaped scarring from beaver bites inflicted during territorial disputes on the underside of a tanned beaver pelt. (R. Needham)

3.4 Habitat and territoriality

Eurasian beavers occupy freshwater bodies (ponds, streams, rivers, marshes and lakes) throughout much of Europe. Populations are still scattered in western and southern regions of this range. In Asia, there are relict populations in China (Xinjiang) and in Mongolia. The species is expanding rapidly in both distribution and numbers (Halley *et al.* 2012). Although they are highly adaptable and can modify many types of natural, cultivated and artificial habitats, they prefer still or slow-moving water with stable depths of at least 60 cm (Gurnell *et al.* 2009). Where these habitats are unavailable or are already colonised by other beavers, they will colonise narrower watercourses and construct dams to create suitable habitat.

Although the average size of a beaver territory is approximately 3 km of shore length, size can vary greatly depending on habitat quality, particularly winter food resources, from 0.5 to 20 km of shore length (Macdonald *et al.* 1995; Herr and Rosell 2004; Campbell *et al.* 2005). During the Scottish Beaver Trial the total length of waters' edge used by each beaver pair or family ranged from 1.8 to 4.7 km (Harrington *et al.* 2015). The extent

Figure 3.5 Newly created wetland through beaver dam-building activity at the Devon Wildlife Trial site. (D. Plummer)

and character of a territory is also affected by the surrounding density of the beaver population.

It is a common misconception that suitable beaver habitat solely comprises large tracts of wet woodland. In developed landscapes, they will readily exploit any palatable vegetation in close proximity to water bodies such as mown amenity bankings, grass verges, grazing pastures or agricultural crops. In climates that experience prolonged winters below freezing, access to woody browse in order to collect a cached food reserve is, however, an essential, limiting factor in survival and population growth.

3.5 Diet and feeding

Beavers are entirely herbivorous and will readily consume a wide range of bark, shoots and leaves of woody plants (majority broadleaf species), as well as herbaceous and aquatic vegetation. During spring and summer, up to 90% of their diet is composed of terrestrial, semi-emergent and aquatic species of plants (Nolet *et al.* 1995), though it can be under 30% in Norway (Campbell *et al.* 2013). Although beavers can fell quite large trees (>1 m in diameter), they tend to favour smaller saplings (<5 cm diameter) in order to obtain their bark, side branches and leafy stems (Haarberg and Rosell 2006). Favoured woody plant species include aspen and poplar (*Populus* spp.), willow (*Salix* spp.) and rowan (*Sorbus aucuparia*), while alder (*Alnus glutinosa*) is generally avoided (Fustec *et al.* 2001; Haarberg and Rosell 2003; Iason *et al.* 2014). Foraging generally takes place close to the bank; in Denmark, for example, 95% of beaver-cut stems were within 5 m of water (Elmeros *et al.* 2003); in Norway, 70% of cut stems were within 10 m and 90% within 20 m (Haarberg and Rosell 2006); in Russia, 90% of cut stems were within 13 m of water and 99% within 20 m (Baskin and Sjöberg 2003). Similarly, findings from the Scottish Beaver Trial determined that most foraging activity on trees occurred within 10 m from the waters' edge (Iason *et al.* 2014).

Beavers will forage a few hundred metres away from water to obtain preferred forage species such as aspen or poplar where these are not available in the vicinity of the water's edge. In flatter landscapes, beavers will create or extend their dams to flood the surrounding land in order to access desirable foraging sites. In landscapes with expanding beaver populations, competition for territories means that beavers will utilise faster-flowing watercourses on steeper gradients, though not normally above a 2.5%

Figure 3.6 Beaver feeding on browse at the water's edge. (I. Sargent)

incline (Howard and Larson 1985; Webb *et al*. 1997; Schulte 1998). Some of the territories they establish in these low-quality environments may only be tenable for a single season when the woody resource they depend upon becomes depleted and the small dams built on fast-flowing streams are swept away by the spring floods.

3.6   Behaviours

Beavers spend large amounts of time in fresh water foraging and maintaining their territories, which they mark with castoreum and anal gland secretion (Rosell *et al*. 1998). They defend these established territories as a family against other beavers (Wilsson 1971; Rosell and Thomsen 2006). They do not hibernate, but reduce their activities and establish underwater caches of woody vegetation in front of occupied lodges during autumn to tide them through the winter months (Wilsson 1971; Hartman and Axelsson 2004). Food caches have been observed in Britain, within the Scottish Beaver Trial and Tayside beaver populations (Campbell *et al*. 2012; Harrington *et al*. 2014). The material in caches which remains unused can, on occasion, rapidly regenerate by sending out new side shoots, particularly where it contains willow. Beavers are strong and able diggers and can excavate burrows, chambers and canals (Wilsson 1971; Richard 1967). Earth, sticks and branches, stones and vegetation are used for construction purposes. Although beavers often live in purpose-built lodges constructed from compacted mud and cut branches, they may also utilise a number of burrows in the same territory (Campbell *et al*. 2012a). Where the banks of a watercourse allow the construction of large burrows, lodge-building is uncommon. Burrows can be re-roofed with sticks and mud where sections collapse.

Figure 3.7 Beaver lodge with extended food cache (branches in water to left of main structure), Bavaria. (R. Campbell-Palmer)

Next to tree-felling, beavers are perhaps best known for their dam-building behaviour. Beavers normally only build dams when necessary to retain and manage water levels. For example, no dams exist on any of the main river channels in an area in Telemark, Norway, in which several of the authors have spent time investigating the behaviour and ecology of beavers. The creation of dams by beavers depends on habitat characteristics. On lakes or wide rivers (>10 m), damming is largely unknown. Beavers living on narrower water bodies (<6 m wide and in 97% of cases in water originally <0.7 m deep: Hartman and Törnlöv 2006) often build dams, and can create extensive systems of multiple impoundments. It appears that a depth of ~70 cm behind the dam is the 'target', though higher dams have been observed, especially on streams with high banks. A 2012 survey across the Tayside catchment found 7 dams with an average height of 0.75cm, with 32 dams (majority short-lived) then recorded between September 2013 and November 2014 (TBSG 2015). The incidence of damming varies according to the characteristics of a watershed. On one river system in Norway, ~10% of the beaver territories had actively maintained dams on river tributaries (Parker and Rønning 2007); in a Polish study, 19.5% of beaver territories had active damming (Żurowski and Kasperczyk 1986), while Russian studies have reported figures ranging from 19 to 53% (Danilov and Kan'shiev 1983).

Dams provide deeper water and a safer environment for beavers to move through, and create deeper and constant water levels concealing the entrance of their main lodge. Beavers strongly prefer to move through water as opposed to travelling overland, which may involve more energy expenditure and leaves them vulnerable to predators. Where they do travel overland on a regular basis, they prefer to use short routes. Over time, these features, for example across a narrow banking separating one watercourse from another, are commonly excavated to create canals. The deeper water retained by a dammed beaver pond affords submerged access to food caches should the surface become frozen in winter. The length of time that dams persist in the environment varies and can be relatively short, particularly if the food resources being utilised become depleted or the dams are not worth maintaining (Halley *et al.* 2009; Rosell *et al.* 2005).

Figure 3.8 Typical beaver dam on narrow watercourse, Scotland. (D. Gow)

A few modern records, all from North America, exist of the collapse of beaver dams resulting in downstream damage to land or travel infrastructure (Butler and Malanson 2005). The materials released by most breached dams are most commonly trapped elsewhere in the watercourse further downstream. Beaver dams rarely collapse as entire structures, but rather breach at a limited point which then erodes through water action. If beavers are still present at the site, they will maintain the dam and rapidly repair any breaches. Such breaches typically occur in autumn and spring at high water discharge (Halley *et al.* 2009).

Beavers are renowned for using their tails to generate alarm signals to warn other family members of danger, but additionally let potential predators know they have been spotted. 'Tail-slapping', as the name suggests, involves a beaver raising its tail above the water surface, then bringing it down sharply to slap the water's surface. Often the beaver dives immediately afterwards. Like most social mammals, interactions between individuals serve to reinforce family bonds. They are more common in younger animals and tend to decline with age. Beavers can produce a range of vocalisations; and family members will commonly produce whining calls when they meet, with young animals tending to be much more vocal. Juveniles will communicate with a range of mews, short, soft squeaking calls or repetitive crying noises, to which the adults respond (Wilsson 1971). Territorial display behaviours, involving the repeated lifting of a cut branch or other object, have been reported in the well-studied Norwegian populations (Thomsen *et al.* 2006) and likely occur elsewhere.

Figure 3.9 Tail-slapping in beavers. (Rachael Campbell-Palmer)

Grooming is an important behaviour which maintains the beavers' fur in good condition and removes parasites. Beavers have two distinct layers of fur: the soft under-fur, which is very dense; and the outer, coarser guard-hair layer. This structure ensures that, when they submerge in water, a layer of air is trapped next to their skin which helps to repel water and offers very effective insulation properties in cold climates. This prevents saturation and allows beavers to dry themselves quickly when they emerge from the water. Beavers can submerge for up to 15 minutes when they feel threatened or attempt to avoid predators; but most dives are to collect aquatic plants, are much shorter in duration (a few minutes), and are repeated frequently when foraging (Novak 1987).

3.7 Parasites and diseases

The range of pathogens that can be harboured by the Eurasian beaver has been reviewed and health-screening recommendations made for importation of beavers to Scotland (Goodman *et al*. 2012; Campbell-Palmer *et al*. 2015). Beavers can carry host-specific parasites not currently present in Britain, though these are not known to infect or harm other species. These include the beaver beetle *Platypsyllus castoris*, a stomach nematode *Travassosius rufus*, and a specialised trematode or intestinal fluke *Stichorchis subtriquetrus*. These species have now all been recorded in wild beavers in Scotland (Campbell-Palmer *et al*. 2012; Goodman *et al*. 2012; Duff *et al*. 2013). Non-native, host-specific parasites are not of concern to human, livestock or other wildlife health, so no active management for these species is presumed to be required. Other parasites such as *Giardia* spp. and *Cryptosporidium* spp. are already present in British wildlife, livestock and humans, and therefore it is possible beavers may contract them and, like other species, contaminate raw water sources. A risk assessment undertaken by the Centre of Expertise on Animal Disease Outbreaks determined that these other sources of infection (humans, livestock and other wildlife) are likely to pose a more significant risk to water contamination, though they recommend appropriate risk assessment for future beaver releases. In conclusion, the likelihood of beavers being important sources of contamination for these two parasites is considered unlikely (EPIC 2015). There are no published reports of *Mycobacterium bovis* infections in the Eurasian beaver and therefore it is not viewed as a major risk factor for domestic livestock. However, as with any other mammal, beavers could theoretically become infected with bovine TB if exposed to an animal actively shedding the organism.

Any current beaver population of unknown origin may carry non-native diseases and parasites, in particular *Echinococcus multilocularis* (also known as the fox tapeworm), rabies and tularaemia. Any health-screening programme involving beavers should include the sampling of any cadavers and/or a live-trapping programme to collect blood and faecal samples as required. *E. multilocularis* is a zoonotic parasite of serious health concern, and is regarded as one of the most pathogenic parasitic zoonoses in the northern hemisphere (North America, northern and central Eurasia) (Eckert *et al.* 2000; Vuitton *et al.* 2003). Although it is established in many countries across Central Europe, other European countries are presently deemed free of this parasite – including the United Kingdom, which employs strict measures to prevent entry, i.e. the Pet Travel Scheme (DEFRA 2012).

Barlow *et al.* (2011) diagnosed *E. multilocularis* in a captive beaver at post-mortem. This individual was held in an English captive collection but had been directly wild-caught and imported from Bavaria (in 2007), Germany. Sample screening across the Tay and Earn catchments, and ongoing post-mortem examination of beaver cadavers, has demonstrated no evidence of *E. multilocularis* in free-living beavers in Scotland. The rabies virus has not been reported in Eurasian beavers but theoretically may affect any mammal. Screening of the live animal is not currently possible, so any imported beavers should be sourced from rabies-free areas or quarantined according to the current Rabies Importation Order (as amended). Like all other rodents, beavers may harbour common European rodent pathogens (Goodman *et al.* 2012). It is recommended that any imported, wild-caught beavers are screened for the following as a minimum: *E. multilocularis*, hantavirus, tularaemia, *Yersinia* spp., leptospirosis, *Salmonella* spp., *Campylobacter* spp. and *Toxoplasma gondii*, and quarantined for rabies.

3.8   Population biology

Beaver survival, population establishment, development and distribution in Britain will directly impact on the management requirements of this species. At low densities, beavers have the ability to blend unobtrusively into an environment, with any conflicts tending to be localised. During this initial phase of colonisation, they select the most favourable sites, typically larger rivers and lakes, where dam-building activity is rare. As beaver populations grow and the population density increases, successive generations occupy less-favoured habitats (i.e. those more likely to be modified by beavers) in minor watercourses or anthropogenic environments. In such locations, their presence can become more obvious as environments are modified, often through a process of dam creation to increase water levels for protection of natal lodges and access to food resources, and often with more obvious feeding impacts. It is generally at this point that conflicts with human land-use interests become more likely. Dam creation and its attendant landscape alteration is the most common cause of conflicts.

Findings from the Scottish Beaver Trial, the Tayside Beaver Study Group, River Otter and various reports on both free-living beavers and enclosure projects throughout Britain clearly demonstrate that Eurasian beavers can survive and adapt well to the British landscape (Campbell *et al.* 2012a; Campbell-Palmer *et al.* 2015; Harrington *et al.* 2015; TBSG 2015). Although their established predators, such as the European Wolf (*Canis lupus lupus*), European lynx (*Lynx lynx*) and brown bear (*Ursus arctos*), are now absent from Britain, young beavers and particularly kits can still be predated by red fox (*Vulpes vulpes*), domestic dogs (*Canis lupus familiaris*), pine marten (*Martes martes*), birds of prey and even large pike (*Esox lucius*) (Kile *et al.* 1996; Rosell and Hovde 1998; Rosell *et al.* 2005). There are anecdotal reports that otters (*Lutra lutra*), American mink (*Neovison vison*) and badgers (*Meles meles*) may also be opportunistic predators, particularly of lone kits.

As beaver populations increase in density, beavers will often be wounded or even killed in territorial conflicts with non-related individuals. Mortality rates across Europe vary greatly depending on a range of factors such as habitat quality, population density, varying climatic conditions, extremes of weather (particularly spring flooding), hunting and trapping, predators (particularly wolves) and disease (Novak 1987; Rosell *et al.* 2005). Many studies demonstrate that mortality rates are highest in the first two years of life (e.g. Payne 1984; Campbell *et al.* 2012b), though survival during the first year after birth can be very high (e.g. 92% vs. 87% for breeding adults, Campbell *et al.* 2012b). Deaths due to road accidents may also be a significant cause of mortality, with small numbers of beaver fatalities already reported on roads in Perthshire (H. Dickinson, personal communication 2014; Campbell-Palmer *et al.* 2015). Other causes of mortality in Europe include flooding and associated drowning (mainly kits), and an inability to wean fully and to cope with dietary changes to vegetation. While captive beavers have been recorded as surviving to 28 years of age, the mean adult life expectancy in the wild is believed to be 12–14 years (Nolet *et al.* 1997), though older territory-holding and reproducing individuals have been recorded in Norwegian study populations (Campbell *et al.* in press).

As with all wildlife, beaver population density varies considerably in time and space and with habitat quality. In addition, beaver density is influenced by their territorial behaviour, with mean territory size tending to decline as habitat quality increases (Parker and Rosell 2012). Since beavers rarely move more than 60 m from water (Barnes and Dibble 1988; Donkor and Fryxell 1999), and most activity is within 20 m of the shore, they occupy landscapes along banksides in a linear fashion. Most landscapes include lengths of unsuitable beaver habitat, e.g. stretches of rapids. Therefore, at the landscape scale, the distribution of beaver territories is often highly discontinuous (Parker *et al.* 2001a). Growth rates in newly established beaver populations are often initially slow, as individuals disperse into a large river system with relatively few potential mates, reducing the rate at which they are encountered. As dispersing offspring may travel dozens of kilometres from their family territories, the process of population establishment creates a 'patchwork' pattern of beaver territories. This can produce a 'lag phase' of slow growth after reintroduction or colonisation of an area, which is then followed by rapid expansion. The length of time required for rapid population expansion varies depending on the characteristics of the river system and may take 15–20 years on larger river systems (Hartman 1995). In release projects, if large (~40+) numbers of animals are released, the chances that dispersing offspring will meet each other sooner are greater and therefore initial population growth rates usually higher.

After this phase of population growth, a decrease in territorial sizes may become evident (Campbell *et al.* 2005). At this point, the availability of habitat becomes a limiting factor, and territorial disputes become more common. Mortality rates increase directly through fighting and indirectly through the stresses of living in smaller territories that need to be defended more vigorously. This whole process can become physically evident by a decrease in breeding rates, delayed dispersal and lighter individual body weights (Busher *et al.* 1983). While a developing beaver population with abundant habitat can display growth rates of 15–20% per annum (Hartman 1995), populations will level off with no demonstrable growth once their whole environment is occupied. However, as long as there is suitable unused habitat available, beaver populations will grow and expand.

Another aspect of beaver ecology that exerts considerable influence on the spatial and temporal distribution, and density of occupied territories, is the alternating pattern of site occupation and abandonment in lower-quality habitat which emerges in mature populations. Following a period of occupation, sites may be abandoned for a number of years if food resources become depleted. Once preferred food species have grown back in sufficient quantity, a new period of colonisation of the site will occur. Thus, a dynamic source-sink pattern of site occupation and abandonment becomes established, with rotation times varying depending on habitat quality and harvest levels (Fryxell 2001). In these populations, measured at a landscape scale large enough to include both good and poor beaver habitats (e.g. >~100 km²), the proportion of sites occupied at any time, i.e. the site occupation rate, tends to vary between 0.33 and 0.50. Thus, after levelling off at an initial population peak, populations may decline somewhat before roughly stabilising. In such populations, between a third and half of the potential beaver habitat within a larger area will be in use at any particular time (Parker and Rosell 2012).

Finally, it is worth noting that a lack of genetic diversity in the initial population may result in inbreeding depression that could limit the rate of population growth and geographic expansion.

The mean litter size for the Eurasian beaver based on fetus counts is approximately 2.5 and based on young born approximately 2.0 (Danilov and Kan'shiev 1983; Mörner 1990; Danilov *et al.* 2011; Parker *et al.* 2012); however, delayed sexual maturity beyond the age of 3 years is common, and sexually mature females may not breed in all years, a reproductive response apparently linked to persistently high densities and dwindling food supplies (Campbell 2010). In southern Scandinavian landscapes ≥100 km² in area and encompassing beaver habitats of varying quality, the mean density of occupied sites in four studies was 0.26/km² (range = 0.32–0.20). On a smaller scale, the highest densities tended to occur in agricultural landscapes with a prevalence of low-gradient streams (Parker *et al.* 2013).

In management terms, after reintroduction, slow establishment may therefore be expected, followed by a period of rapid growth and dispersal, before becoming more stable at a lower population size and density (Figure 3.10). Culling both during the period of rapid growth or rough stability following is likely to be followed by rapid recolonisation by surplus animals. Once beaver territories become fully occupied, non-lethal management, which permits beaver presence within tolerable limits, should slow or prevent further colonisation and significant increase in density due to the territoriality of this species. In reintroduced populations in areas of Sweden and Norway, it has taken several decades for beavers to attain typical population densities and start to occupy suboptimal habitats (Hartman 1994; Halley 1995). It is typically not until this stage that a higher incidence of human–wildlife conflict occurs (Bhat *et al.* 1993; Deblinger *et al.* 1999).

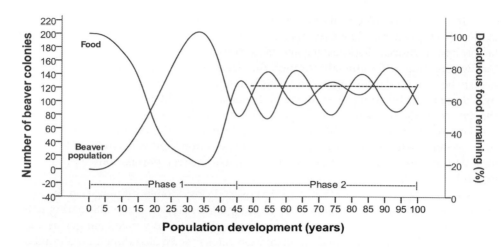

Population development (years)

Figure 3.10 A graphical presentation of the predicted relationship between a developing beaver population and its prime winter food source of deciduous trees and bushes. Assumptions for the model are an area of 500 km² of typical beaver habitat in Fennoscandian boreal forest with a peak in colony density occurring at 0.40 active territories/km² after 35 years, and a mean active territory density of 0.25/km² (dotted black line) following population recovery. Culling pressure is light or non-existent. The basic shape of the curve for colony number in 'phase 1' is relatively well founded in field studies. The shape in 'phase 2' is based on limited anecdotal observation. (Parker and Rosell 2012)

3.9   North American beaver introductions

Since their release, North American beavers have spread throughout southwest and central Finland and into Karelian Russia. They have established a presence in parts of Belgium, western Germany and Luxembourg as a result of zoo escapes (Dewas *et al.* 2012).

Expanding populations of both species have converged on two fronts in Finland and northwest Russia (Parker *et al.* 2012). The body size of both species is similar, though the litter size of the North American beaver is slightly larger. Only minor differences in life history, ecology and behaviour exist, suggesting a nearly complete niche overlap. The question is whether coexistence or competitive exclusion will ultimately result, with the possible regional extirpation or eventual extinction of one of the species. This would be of concern if the outcompeted species was the Eurasian beaver. Though extirpation of North American beaver from continental Eurasia is still possible (Parker *et al.* 2012), it may never happen given a lack of willingness to accept the seriousness of the problem and limited resources for wildlife management.

Preventative measures incorporated in the beaver-management plan for the Czech Republic include a database of all the captive collections holding North American beavers, the establishment of cooperation to ensure prompt attempts to recapture any escaped animals, the elimination of any identified wild-living North American beaver as a non-native species, and the promotion to captive collections of the replacement of North American individuals with Eurasian Beavers (Vorel *et al.* 2013). Given this experience and the documented ability of beavers to escape from captivity, it would be prudent to reduce the potential for introduction of North American beavers to Britain.

Very few captive facilities in Britain currently hold North American beavers. The species can, however, be readily imported. Greater consideration should be given to the purpose of such animal movements, to sterilisation, or to replacement with Eurasian beavers. The registration and permanent individual identification of any captive North American beavers in Britain would be sensible.

Key concepts

- There are two species of beaver – Eurasian and North American/Canadian – which can be difficult to tell apart in the field. North American beavers have been introduced or have escaped in parts of Europe, presenting a management issue. Preventative action should ensure this does not occur in Britain. Currently, there are no known wild-living North American/Canadian beavers in Britain.
- Beavers are large (20 kg+), semi-aquatic, completely herbivorous rodents.
- Sexing beavers without handling is difficult unless the nipples of heavily pregnant or lactating females are visible. This is because both sexes conceal their sexual organs within a cloaca. With experience and appropriate handling beavers can be sexed according to the colour and viscosity of their anal gland secretions.
- Beavers live in family units, with one dominant pair of breeding adults and their offspring from 1–2 breeding seasons. They produce one litter of typically 2–4 kits per year. Families are highly territorial.
- In the wild beaver mortality can be high during the first 2 years of life, but if they survive this period they can live on average to 12–14 years.
- Beavers live in freshwater habitats, with the majority of their on-land activity occurring within 20 m of the shoreline.
- They are subject to the same range of parasites and diseases as other wildlife, though directly imported animals should be screened for non-native pathogens.
- Initial population growth on a river system is typically relatively slow, then rapid for a period. As long as there is available suitable unoccupied habitat, beaver populations will grow and expand.
- Populations then peak and may decline somewhat as lower quality habitat which cannot sustain beavers permanently goes out of use for a period while the vegetation recovers. Thereafter populations reach rough stability, regulated by the territorial nature of beavers, and with lower quality sites cycling in and out of use.
- Kits in Britain will be subject to predation pressure, and roadkill may be a significant cause of mortality for dispersers and adults.

4. Legislation

The Eurasian beaver is protected across Europe through a number of legislative measures. The framework for protection is provided by the Convention on the Conservation of European Wildlife and Natural Habitats, and specific provisions are made by European Union Directive 92/43/EEC on the Conservation of Natural Habitats and of Wild Fauna and Flora (the 'Habitats Directive'; Species of Community Interest in Annex II and Annex IV(a)) (http://ec.europa.eu/environment/nature/legislation/habitatsdirective/index_en.htm). In EU member states (with some specific exceptions), wild beavers present within their natural range are protected from deliberate capture, killing and/or disturbance, and damage or destruction of their breeding sites or resting places. It is also prohibited to keep, transport or offer for exchange or sale specimens taken from the wild. Additional protection of beaver habitats is afforded through other legislation such as the Ramsar Convention and by the Natura 2000 network.

The Habitats Directive is transposed into domestic legislation in England and Wales by the Conservation of Habitats and Species Regulations 2010 and in Scotland by the Conservation (Natural Habitats, &c.) Regulations 1994 (as amended). Table 4.1 summarises the main legislation that would apply to the Eurasian beaver in Britain. Under domestic law, it is an offence to keep beavers taken from the wild (unless it can be shown the beavers originated from outside the EU), except under licence by the relevant authority (Natural England, Scottish Natural Heritage or Natural Resources Wales), unless it can be proven that they are lawfully held, having been taken from the wild in an EU member state without contravention of the law and before the implementation date of the Directive (10 June 1994 for most member states, later for some accession countries). Captive-bred offspring are only considered such (and therefore exempt from this requirement) if their parents were lawfully held in captivity. However, the beaver is not currently listed in the Regulations as a 'European Protected Species', therefore, at the time of writing, the full protection afforded by the Directive is not transposed into domestic law. Formal reintroduction, for example, following the successful Scottish Beaver Trial in Scotland, is likely to require the addition of beavers to the Regulations, thus affording them full protection.

In England and Wales, section 14 of the Wildlife and Countryside Act (1981) prohibits the release of any animal not ordinarily resident, or a regular visitor in Great Britain in a wild state, or which is listed in Schedule 9. The Eurasian beaver was recently listed in Schedule 9, Part 1B ('Animals no longer normally present'; amended by the Infrastructure Act 2015), so their release, without a licence, is prohibited. In Scotland, section 14 of the Wildlife and Countryside Act (1981) (as amended) prohibits the release, without a licence, of any animal outside of its native range. Scottish Natural Heritage currently considers beavers to be outside of their 'native range' when in Scotland, as defined under the Wildlife and Natural Environment (Scotland) Act 2011 and the Habitats Directive.

Beavers, like all wild mammals, are afforded limited protection against cruel acts, or unnecessary suffering, by the Wild Mammals (Protection) Act 1996, animal-welfare legislation and hunting legislation.

Due to expansion in beaver populations and distribution in Eurasia in recent decades, the beaver was recently downgraded on the International Union for Conservation of Nature (IUCN) Red List of threatened species to 'Least Concern', the most favourable conservation status. However, conservation measures are still recommended, including closed hunting seasons, temporary or local bans on particular exploitations, and the regulation of trade in live or dead specimens, in order to ensure that the species does not once again become endangered (Batbold *et al.* 2008, http://conventions.coe.int/Treaty/EN/Treaties/Html/104.htm). This downgrading does not affect the status of Eurasian beavers under the Bern Convention or the Habitats Directive.

Key concepts

- The Eurasian beaver is listed in the Habitats Directive as a Species of Community Interest, though several countries have reduced protection status, allowing management options such as regulated hunting while maintaining favourable conservation status.
- Currently, a licence is required under the Wildlife & Countryside Act, to release, possess or transport beavers in Britain.
- Future legislation is likely to change and potentially differ between Scotland, England and Wales, and therefore the relevant Statutory Nature Conservation Organisations should be consulted before undertaking any beaver management.

Table 4.1 Current legislation most relevant to Eurasian beavers in Britain (modified from SNH 2015, with additional information provided by Natural England). The authors would recommend seeking legal advice when necessary and checking with the relevant statutory authority.

Legislation (as amended where applicable)	Summary description	Relevance to beavers
Nature-conservation law		
EU Directive 92/43/EEC on the Conservation of Natural Habitats and of Wild Fauna and Flora (the 'Habitats Directive')	The Directive requires member states to take appropriate measures for the strict protection of Species of Community Interest (listed in Annex IVa of the Directive) and, for species in Annex II, to designate Special Areas of Conservation (SACs). It also requires member states to study the desirability of reintroducing Annex IVa species that are native to their territory where this might contribute to their conservation.	Beavers are listed in Annex II and Annex IVa (some EU populations are specifically exempted).
The Conservation (Natural Habitats, &c.) Regulations 1994 (as amended) *Scotland only*	The Regulations implement the provisions of the Habitats Directive in Scotland and describe the protection given to European Protected Species (EPS; those Annex IVa animals whose natural range includes any area of Great Britain), and the licensing regime applicable to those species. The Regulations also set out the Natura site designation process and assessment implications. (Natura sites being areas of international importance designated as Special Areas of Conservation or Special Protection areas)	Very limited protection given to beaver at the moment throughout Britain. If formally reintroduced, beavers are likely to be added to the European Protected Species list (EPS; those Annex Iva animals whose natural range includes any area of Great Britain) and require strict protection (this can be done for Scotland only within the UK). A licensing regime would then have to be put in place to allow for management actions that conflict with this protection. Site(s) may require designation as SACs (Special Areas of Conservation) for beavers. Plans or proposals affecting beaver SACs would require assessment in the light of the site's conservation objectives before being approved. Plans or proposals affecting any Natura site (SAC or Special Protection Area for birds), including any beaver reintroduction, would also require a 'Habitats Regulations Appraisal' before proceeding. Some of these might require an 'Appropriate Assessment' before a decision is made about whether or not to proceed.

Table 4.1 – *continued*

Legislation (as amended where applicable)	Summary description	Relevance to beavers
The Conservation of Habitats and Species Regulations 2010 (as amended) *England and Wales only*	The Regulations implement the provisions of the Habitats Directive in England and Wales and describe the protection given to European Protected Species (EPS; those Annex IVa animals whose natural range includes any area of Great Britain), and the licensing regime applicable to those species. The Regulations also set out the Natura site designation process and assessment implications.	Very limited protection given to beaver at the moment throughout Britain. Under Regulation 41(3), it is unlawful to possess or transport any beaver that was taken from the wild expect under licence. This applies to living and dead beavers, and to any part or derivative of a beaver (e.g. pelts). If the decision in Scotland, following the successful Scottish Beaver Trial, is to formally recognise the beaver as a reintroduced species, they are likely to be added to the EPS list in the Regulations and require strict protection. A licensing regime would then have to be put in place to allow for management actions that conflict with this protection. Following the listing of Eurasian beaver on Schedule 9 of the Wildlife & Countryside Act in 2015 (Infrastructure Act 2015), the Government is considering putting the beaver onto Schedule 2 of the Regulations (European Protected Species of Animals). Site(s) may require designation as SACs for beavers. Plans or proposals affecting beaver SACs would require assessment in the light of the site's conservation objectives before being approved. Plans or proposals affecting any Natura site (SAC or Special Protection Area for birds), including any beaver reintroduction, would also require a 'Habitats Regulations Appraisal' before proceeding. Some of these might require an 'Appropriate Assessment' before a decision is made about whether or not to proceed.

Table 4.1 – *continued*

Legislation (as amended where applicable)	Summary description	Relevance to beavers
Wildlife and Countryside Act 1981 (as amended) (release of non-native or former native species) Wildlife & Countryside Act 1981 (as amended) (prohibition of certain methods of killing or taking wild animals)	Under section 14 of this Act, it is illegal to release or allow to escape into the wild any species that is not ordinarily resident or a regular visitor to Great Britain in a wild state or is listed in the relevant part of Schedule 9 of the Act (England and Wales) or, in Scotland, to release a species outside its native range (as defined in the Act) or specified by an Order of the Scottish Ministers, without an appropriate licence. Under section 11 of this Act, the use of certain methods of taking or killing are prohibited for animals listed on Schedule 6	The Eurasian beaver is listed on Schedule 9(1)(b), 'Native animals no longer present' of the Wildlife & Countryside Act 1981. A licence is required to release beavers into the wild. Any release of beaver into the wild would require a licence from the relevant statutory nature conservation body. Eurasian beaver will be added to Schedule 6 in 2016 prior to the implementation of the Agreement on International Humane Trapping Standards (AIHTS) in the UK in July 2016.
Nature Conservation (Scotland) Act 2004 / Wildlife and Countryside Act 1981 (as amended) (impact on protected sites)	Under the relevant legislation, the statutory nature conservation bodies notify Sites of Special Scientific Interest (SSSIs) and describe 'operations requiring consent' (ORCs) or 'operations likely to damage' (OLDs) the features of interest. Acts or omissions which might damage features of interest require consent before being carried out.	If released onto an existing SSSI, beaver release or subsequent management might require consent.
Salmon and Freshwater Fisheries (Consolidation) (Scotland) Act 2003 / Salmon and Freshwater Fisheries Act 1975	These set out the law and regulatory areas concerning freshwater fisheries, including offences in relation to passage of salmon or migratory trout. Obstructing passage of salmon or migratory trout may result in an offence.	Implications of possible riverine habitat change/engineering resulting from beaver activity (e.g. dam construction), or beaver management, which might impede fish movement within river systems and affect in-stream habitat.

Table 4.1 – *continued*

Legislation (as amended where applicable)	Summary description	Relevance to beavers
Trade and movement of animals		
EU Directive 92/65/EEC (the 'Balai Directive')/Rabies (Importation of Dogs, Cats & Other Mammals) Order 1974	The Balai Directive provides a framework for animal health requirements governing trade in captive-held animals between EU member states and imports into the EU. There are health certification requirements, and premises holding animals need to be registered and approved by the Animal and Plant Health Agency (APHA). Animals listed under the Rabies Order 1974 are subject to quarantine controls.	All imported beavers held in captivity in the UK should be accompanied by an appropriate health certificate. Premises holding imported beavers need to be registered with the APHA. Beavers are listed on Schedule 1 Part II of the Rabies Order and imports of wild-caught animals are subject to the provisions of the Order including six months' quarantine at an approved establishment (although exceptions have been applied in Scotland for the Scottish Beaver Trial). If beavers are to be exported to an EU country, the exporter requires an export health certificate.
Animal-welfare law		
Animal Health and Welfare (Scotland) Act 2006/Animal Welfare Act 2006	These Acts are primarily concerned with domestic animals but their provisions also apply to wild animals if they are under the control of man, whether on a temporary or permanent basis, or they are not living in a wild state. Such animals are protected against action, or failure to take action, that results in unnecessary suffering. In the UK animals transported by air (either outside or within the UK) must comply with the International Air Transport Association's 'Live Animals Regulations' (LAR).	Beaver welfare should be considered when animals are captured, transported or held in captivity, and during and after release into the wild. Persons responsible for holding beavers in captivity must not cause them unnecessary suffering or fail to take reasonable steps to ensure their welfare. Where capture or release of beavers is undertaken in another country, the relevant animal welfare legislation of that country must be adhered to.
European Zoos Directive 1999 Zoo Licensing Act 1981	Establishments holding wild animals kept for exhibition (other than circuses or pet shops), where the public have access for 7 or more days a year, must be inspected and licensed in most cases. Under the 1981 Act, persons wishing to operate a zoo must be licensed by the local authority.	Establishments holding beavers for exhibition purposes and open to the public for at least 7 days a year must be inspected and licensed by the relevant local authority.

Table 4.1 – *continued*

Legislation (as amended where applicable)	Summary description	Relevance to beavers
Wild Mammals (Protection) Act 1996	This Act protects mammals living in the wild from cruel acts (e.g. stoning, impaling, burning, drowning, crushing, dragging) carried out with the intention to inflict unnecessary suffering.	Beavers living in the wild (whether legally or illegally released) are protected by this legislation.
Water and flood-risk management		
EU Water Framework Directive 2000/60/EC Water Environment and Water Services (Scotland) Act 2003 *The Water Environment (Water Framework Directive) (England and Wales) Regulations* 2003 Water Environment (Controlled Activities) (Scotland) Regulations 2011 ('CAR')	The Directive establishes a framework for the protection of inland waters, estuaries, coastal waters and ground water. The 2003 Acts transpose the Directive into UK law with respect to river basins and other elements of the water environment whilst the 2011 Regulations in Scotland give Scottish Ministers regulatory controls over water activities. Persons intending to carry out any activity which might affect the water environment require authorisation from the Scottish Environmental Protection Authority (SEPA) or the Environmental Agency (EA).	The management of beavers on a site might require consultation with the EA/SEPA and may require a CAR application to SEPA (e.g. river impoundment works to protect river banks). SEPA has developed a pragmatic position statement on the management of beaver structures in Scotland (available from the SEPA website).
EU Directive 2007/60/EC on the assessment and management of flood risks. Flood Risk Management (Scotland) Act 2009 Flood and Water Management Act 2010	The Directive aims to establish a framework of measures to reduce the risks of flood damage. The 2009 and 2010 Acts transpose the Directive into UK law, introducing requirements to reduce the adverse consequences of flooding for a range of reasons, including human health and the environment. They set out areas of responsibility for assessing and managing flooding and place a strong emphasis on working with nature to manage flood risk.	Habitat change brought about by beaver activity might contribute to restoring natural processes within catchments. Beaver presence might increase or reduce flood risk at a local level. Strategic and local flood-risk management planning will need to take account of potential beaver activity in managing flood risk sustainably.

Table 4.1 – *continued*

Legislation (as amended where applicable)	Summary description	Relevance to beavers
Reservoirs (Scotland) Act 2011 Reservoirs Act 1975, as amended by the Flood and Water Management Act 2010	These set down the regulatory regime for the safe construction and operation of 'controlled reservoirs'. They require registration of controlled reservoirs, regulate their construction and specify inspection requirements. SEPA/EA must assess the risk of uncontrolled releases of water from controlled reservoirs (in terms of adverse consequences and probability). They also give SEPA/EA the power to act in an emergency to protect people or property from water escaping from a controlled reservoir.	There is the potential for beaver burrowing, for example, to damage 'controlled reservoirs' with consequent risk to public safety and infrastructure. More frequent inspection of some controlled reservoirs may be required. Plans for new reservoirs might need to take into account possible beaver activity in the area.
Environmental liability and impact assessments Environmental Liability Directive 2004 The Environmental Damage (Prevention and Remediation) (England) Regulations 2015 The Environmental Damage (Prevention and Remediation) (Wales) Regulations 2009 (as amended)	Under the Directive and the transposed UK laws, operators causing, or causing a risk of, environmental damage (which includes offences affecting Annex II and Annex IV species and their breeding sites or resting places) are held financially liable for remedying the damage. Protection applies whether the species is inside or outside a Natura site.	Operators who impact the conservation status of beavers (e.g. by killing animals or damaging their breeding sites or resting places) may be held liable for remedying the situation.
EU Directive 2001/42/EC Strategic Environmental Assessment Directive Various domestic laws and regulations	This Directive, and the domestic legislation transposing it into UK law, require that any public body preparing certain plans must carry out a strategic environmental assessment of them if they are likely to have significant environmental effects.	Relevant authorities will need to consider whether or not the reintroduction of beaver in a particular area requires a strategic environmental assessment and, if so, arrange for one to be completed.

5. Effects of beavers

The activity of beavers creates habitats that are dynamic in nature. Beaver activity can provide a wide range of ecological and economic benefits, but it is clear that, in landscapes which have been moulded by people, beaver activity will have to function within limits acceptable to human interests. While a range of practical measures can be employed to mitigate some impacts considered undesirable, any sustainable long-term view of beavers' presence should seek to develop a niche for their existence as the creators and modifiers of multifunctional wetland environments. To integrate beavers into the management of cultural landscapes, identification of conflicts with human interests should occur as soon as possible, and management techniques should be implemented before issues become more widespread and more expensive to rectify. A comprehensive review of the effects of beavers on the natural environment and each major taxonomic group, along with experience in Scotland to date and potential future implications is presented in SNH's 'Beavers in Scotland' report (SNH 2015).

Table 5.1 Summary of the main benefits and disadvantages of restoring beaver populations

Benefits	Costs
• Enhance biodiversity • Create spawning grounds and refuge areas for some fish species • Stabilise water fluctuations and store water • Increase habitat heterogeneity and create dynamic environments, create food and cover for other species • Socioeconomic impacts – income to landowners, ecotourism, agri-environment schemes, payments for water retention • Ecological services – carbon sequestration, sediment and diffuse pollution trapping, improvement in water quality • Favourable growth of riparian plants and stabilisation of riparian banks in some situations • Source of fur, meat, recreational hunting	• Flooding of land and damage to transport and drainage infrastructure • Dams creating potential barriers to fish passage • Damage to agricultural machinery through burrow collapse • Browsing on agricultural crops • Damage to flood banks, fish ponds and infrastructure through burrowing • Deleterious impact on isolated populations of protected species • Damage to specimen trees, garden ponds and amenity riparian features • Foraging on agricultural crops and commercial forestry

5.1   Beavers as ecosystem engineers

Beavers are a keystone species whose niche as wetland engineers has a significant impact on the natural landscape. Their activities can result in the formation of wetland habitats, with a positive effect on plant and animal diversity. A recent meta-analysis determined that overall beavers have a positive effect on biodiversity (Stringer and Gaywood in press). Beaver-influenced environments, with a less dense or intermittent tree canopy, provide both a greater expanse and increased variety of living opportunities for a wide range of higher plant species, which in turn increases the feeding and breeding opportunities for insects. A greater abundance of both standing and submerged dead wood habitat further enhances this process, and invertebrate densities can alter significantly in response (Gurnell *et al.* 2009). Beaver-generated impacts (such as damming and associated pond creation) on aquatic invertebrates can be significant, with gravel-dwelling species (e.g. stoneflies, *Plecoptera* spp.) generally being replaced by silt-dwelling species (e.g. mayfly, *Ephemeroptera* spp. and dragonfly, *Anisoptera* spp. larvae) (Knudsen 1962). It has been demonstrated that beaver-dammed streams in Sweden contain a greater range of invertebrates than those without beavers (Sjöberg 1998). Studies in Germany have shown that beaver territories have significantly greater numbers of dragonflies and damselflies than areas without beavers (Harthun 1999). The presence of beavers thus tends to have a beneficial impact on fish populations through the creation of foraging and shelter habitats for a wide variety of species (Rosell *et al.* 2005; Collen and Gibson 2000).

Beaver ponds, formed behind dams, can increase the densities and species richness of trout or coarse fish communities by providing refugia from low flows and enhanced feeding opportunities (Schlosser 1998; Snodgrass and Meffe 1999). Tree-felling activities result in a greater abundance of submerged woody debris. An associated increase in invertebrate diversity and biomass can result in enhanced growth rates in fish such as

Figure 5.1 Snails feeding on beaver-stripped tree trunk. (R. Campbell-Palmer)

Figure 5.2 Trout caught in a beaver pond, Norway. (D. Halley)

Atlantic salmon (*Salmo salar*) parr (Sigourney *et al.* 2006). The refugia provided by felled trees, woody debris, beaver lodges and food caches has also been demonstrated to attract fish such as salmonids and percids (Collen and Gibson 2000).

Amphibian abundance in beaver-generated landscapes tends to be higher than in areas without beavers (France 1997; Metts *et al.* 2001), and this has also generally been found to be true for reptiles in North America (Metts *et al.* 2001). As British reptiles are all terrestrial (though grass snakes are often found in and around fresh water), the presence of beavers is unlikely to affect them significantly (Gurnell *et al.* 2009). As our knowledge from archaeological evidence of a once-broader herpetofauna continues to develop, it may well be that beaver-generated landscapes could provide sustainable environments for the wider restoration of species such as the pool frog (*Pelophylax lessonae*).

Beaver-generated landscapes have a positive impact on the diversity and density of bird species through the creation of new wetland habitats and enhanced foraging opportunities (Medin 1990; Grover and Baldassarre 1995; Brown *et al.* 1996). Beaver ponds can create successful waterfowl-rearing habitats through increased numbers of invertebrates and the safety of nesting and refuge sites from predators (Nummi and Hahtola 2008). Species such as woodpeckers (*Picidae* spp.) and nuthatch (*Sitta europea*) exploit the standing dead timber, and nocturnal avian predators such as owls (*Strigiformes* spp.) are attracted by increased prey and will utilise abandoned woodpecker nests (Carr 1940; Hilfiker 1991). Mammal species, including bats (*Microchiroptera* spp.) and mustelids (*Mustela* spp.) (Hilfiker 1991) make use of beaver ponds, although non-native species, such as Mandarin duck (*Aix galericulata*) and American mink (*Neovison vison*), may also benefit from beaver-created wetlands. Diurnal avian predators, including ospreys (*Pandion haliaetus*), have been documented using beaver ponds to hunt (Carr 1940; Grover and Baldassarre 1995; Longcore *et al.* 2006). Small mammals (such as water voles *Arvicola amphibius*, water shrews *Neomys fodiens*, or bank voles *Myodes glareolus*) adapt to utilise the variety of niche habitats and prey abundance provided by beaver-generated landscapes (Ulevičius and Janulaitis 2007; Suzuki and McComb 2004), while large herbivores such as deer (*Cervidae* spp.) and European elk (*Alces alces*) exploit their

Figure 5.3 Beaver pond and beaver-generated landscape, providing habitat for numerous plant and animal species. (D. Halley)

greater grazing and browsing potential (Rosell *et al.* 2005). Carnivores, particularly otter, but also potentially badgers (*Meles meles*), red fox (*Vulpes vulpes*), stoats (*Mustela erminea*) and pine martens (*Martes martes*), may benefit through enhanced prey populations (such as fish, amphibians, birds) and the presence of beaver lodges and burrows which they utilise for shelter and breeding (Grasse 1951; Tyurnin 1984; Rosell and Hovde 1998; Rosell *et al.* 2005; LeBlanc *et al.* 2007).

5.2 Beavers and species of high conservation value

In developed cultural landscapes throughout Western Europe, habitat fragmentation as a result of competing land-use pressures poses a challenge to the survival of wildlife and their habitats. In this context, the impact of beavers upon protected areas (e.g. wildlife reserves, national parks, country parks) offers potential benefits and costs. Beaver activity can result in significant changes to plant and animal communities, to the cost of some. While some species are able to exploit these to their benefit, others may be reduced, displaced or even lost (Willby *et al.* 2011). Despite the scientific evidence demonstrating that the *net effect* of beavers is a positive for biodiversity, there may be cases where less mobile species of conservation value could be significantly negatively affected by beaver activity (Stringer and Gaywood in press). It is likely that, under such circumstances, conservation managers could be faced with conflicting management goals. Judgements may have to be made whether to apply conventional mitigation techniques to preserve stability, or to opt for allowing a successional process of environmental change. Practical examples of negative beaver impacts of this type have so far proved to be rare. European experience indicates that, in landscapes which are ecologically impoverished by intensive agriculture, forestry, water engineering or industrial activities, beaver-created habitats are important regenerators of biodiversity.

Concerns regarding the potential of negative impacts on Eurasian aspen (*Populus tremula*) conservation have been expressed in Britain, particularly in relation to important mature stands of high biodiversity value and to sensitive, dependent species such as the aspen hoverfly (*Hammerschmidtia ferruginea*) and a range of Red Data List *Diptera*, lower plants and fungi. Recommendations to promote aspen hoverfly conservation include the protection of fallen aspen/dead wood as a breeding resource, managing sites to promote sustainable quantities of dead wood, and in the long term the reduction of grazing (including fencing) to promote natural regeneration (Malloch Society 2007). The protection by deterrent fencing of key aspen sites may be appropriate, and could be planned and mapped in advance if beavers are reintroduced in Britain. Beavers concentrate their foraging activity close to water (Haarberg and Rosell 2006), therefore aspen sites far (hundreds instead of tens of metres) from water features will experience a reduced risk of felling.

A study has demonstrated that, while beavers can reduce mature tree coverage, beaver felling significantly increases sprouting, and therefore beavers could be utilised to facilitate aspen clone regeneration (as long as ungulate browsing pressure is low) (Runyon *et al.* 2014). Given that the main biodiversity benefits come from larger, mature aspen, beavers would need to be excluded for extended periods of time, as they will tend to fell saplings, thus preventing trees from reaching significant size and maturity. Other tree species known to be preferred by beavers, such as willow, readily coppice, re-root from felled stems or partially debarked cuttings, or grow rapidly from wind-borne seed in damp environments. An increase in stump regeneration in Atlantic hazel (*Corylus avellana*) following increased beaver foraging was recorded at the Scottish Beaver Trial, with re-sprouting increasing from 26% to 52% (Iason *et al.* 2014). Associated lichen species on Atlantic hazel (and aspen) of conservation value may be directly impacted by felling of mature stools and even small gaps in the lichen habitat community may result

Figure 5.4 Beaver-generated flooded forest, Bavaria. (D. Gow)

in lichen loss with the potential of local extinction (Genney 2015). In Britain, the scale of such impacts through beaver activity should be monitored.

Batty (2002) concluded that, overall, beavers would have a beneficial effect on aspen; however, potential conflicts could be reduced by protecting sensitive stands (or single mature trees as appropriate), by not reintroducing beaver into areas prioritised for conservation of stands containing mature aspen trees, but most importantly by increasing the quality (range of sizes and ages) and quantity of aspen currently present. The coexistence of beavers and aspen (both mature trees and smaller shoots) has been documented across Europe. However, in Scotland, the mature aspen resource is extremely limited and fragmented due to a combination of overgrazing and historical land management, and is vulnerable to any new impacts. There is little doubt that beavers can have a significant impact on certain stands (Puplett 2008). Studies undertaken in Scotland within fenced beaver enclosures suggest that rapid regeneration of aspen would occur following significant beaver felling activity in riparian woodlands (Jones *et al.* 2009); but, critically, replacement of mature trees would be slow and may not occur at all, especially if grazing pressure from livestock or deer, for example, is significant.

Concerns have also been raised over potential beaver impacts on specific freshwater species of conservation concern. A knowledge gap concerning beaver interaction with other species including the European eel (*Anguilla anguilla*), lamprey spp. and freshwater pearl mussel (*Margaritifera margaritifera*) has been highlighted by the Beaver Salmon Working Group (BSWG), and would require further monitoring in a Scottish context (BSWG 2015). One study using questionnaires sent to beaver and fish biologists in Austria and Norway suggested that the European eel may benefit from beaver damming activities through increased cover and feeding opportunities (Collen 1997). Overall it is unlikely that there will be any real issues with eel and lamprey and beaver activity in Britain. The freshwater pearl mussel is declining throughout Britain (Cosgrove *et al.* 2000). On some sites in Latvian rivers, populations of pearl mussels have been influenced by beaver dams, which resulted in silt coverage of their gravel beds, temperature changes in the water, shading and eutrophication (Rudzīte 2005; Rudzīte and Znotina 2006). These undesirable changes resulted in the removal of beaver dams and beaver culling

Figure 5.5 Oil trapped behind a beaver dam; water quality downstream of beaver dams has been shown to improve. (D. Gow)

by Nature Conservation Agency staff (Rudzīte and Rudzīte 2011). The justification for this action has, however, been challenged due to the lack of a robust study design with supportive results. Campbell (2006) argues that, since beaver activities can result in more sediment retention, improving water quality downstream and reducing extreme water fluctuations, pearl mussel populations downstream from beaver ponds might actually benefit.

5.3 Beaver effects at a catchment scale

Modern land-use management practices have been demonstrated to increase the generation of surface runoff at a local scale (O'Connell *et al.* 2007). The trapping of sediments (Butler and Malanson 2005) and diffuse pollutants (Cirmo and Driscoll 1993), the dampening of low- and high-flow discharge peaks in watercourses (Beedle 1991; Gurnell 1998), the provision of standing water during times of drought (Westbrook *et al.* 2006) and the potential for carbon sequestration through wetland creation are all potentially economically important aspects of beaver-generated habitats (Wohl 2013). During dry summers, beaver dams and canals have been shown to hold 60% more water (including ground table water) than comparable environments without beaver activity (Hood 2012). On the Keriou River in Brittany, the channel capacity prior to beaver presence was estimated at 535 m^3 over a distance of 120 m. This rose to 3,250 m^3 after a series of beaver dams was constructed (Coles 2006). Studies in Belgium suggest that a series of beaver dams in the upper catchment of the Ardennes has played a significant role in the reduction of discharge peaks and hence flood events in villages lower down in the catchment (Nyssen *et al.* 2001). The slowing of water runoff is also an important function of beaver dams. In one North American study, water was calculated to take 3–4 hours to travel ~2.6 km along a course with no beaver dams, whereas the equivalent water took 11 days to travel the same distance when a beaver dam existed (Müller-Schwarze and Sun 2003). The ecological value of wetlands created by beaver activities is considered to be high but undervalued in terms of water storage, water purification and their ability to regenerate degraded aquatic ecosystems (Czech and Lisle 2003). Beaver recolonisation through watersheds may take a number of years, and therefore these potential benefits may only materialise if beaver activities are allowed to function in a dynamic, flexible manner and on a landscape scale (Gow and Elliott 2014). For example, the landscape-scale influence of beavers is only now being realised in the upper catchments of the River Danube in Bavaria, where beavers were reintroduced in 1966.

In a British context, the concept of beaver activity as potentially providing sustainable flood alleviation is receiving increasing attention (Gow and Elliott 2014). The Pitt Review, a UK Government-commissioned report following the 2007 floods, included a number of measures aimed at slowing down and retaining more water runoff, especially during periods of high and persistent rainfall. This report also made reference to the wider amenity and biodiversity benefits such features could bring (Pitt 2008). Making Space for Water (DEFRA 2005), the Water Framework Directive (WFD 2000/60/EC), DEFRA's Water Strategy (DEFRA 2008), the Environment Agency (2008a, b) and the Flood and Water Management Act 2010 all encourage the restoration of natural processes and the development of sustainable solutions for flood management, particularly in light of climate change (Parrott *et al.* 2009). Belford in Northumberland experienced repeated annual flooding in the years 1999–2009. In 2007, a collaboration between the Environment Agency and the University of Newcastle trialled a series of 'soft' engineering principles such as the creation of storage ponds, wetlands, construction of natural barriers ('beaver dams'), placement of large woody debris (to slow flood peak and divert it onto the flood plain) and replanting of willow along riparian zones within the upper catchment (Wilkinson *et al.* 2010; Nicholson *et al.* 2012). These runoff attenuation features were designed to slow the movement of water, particularly after heavy rainfall. Landowner liaison and buy-in were essential, and the Belford Runoff Management Toolkit provides a guide to implementing such mitigation. The success of this project in terms of increased water storage, improved water quality, slowing of water and sediment runoff, and reduction of subsequent flooding lower in the catchment, along with additional positive biodiversity and ecological benefits, has led to similar initiatives, e.g. 'Slowing the Flow', Vale of Pickering, Yorkshire (Forestry Research 2014).

Probably the most detailed research on the impacts of beavers on hydrology is being carried out by Professor Richard Brazier's team at the University of Exeter. Their instrumentation of the Devon Wildlife Trust Beaver Project site (in which one beaver family has had a significant impact on a small stream running through an area of culm grassland) is investigating the impact of a series of 13 beaver dams along 150 m of watercourses by measuring flows above and below the site. Through damming and other behaviour, the water volume, surface area and ground water of the site have significantly increased (Elliott and Burgess 2013). The volume of water now held within the site is estimated at 650 m³ and flood hydrographs are demonstrating a significant reduction in peak flows and an increase in lag times as flows pass through the now structurally complex site. An important aspect of this research examines how watercourses respond to repeated rainfall events when saturated conditions in upper catchments might otherwise

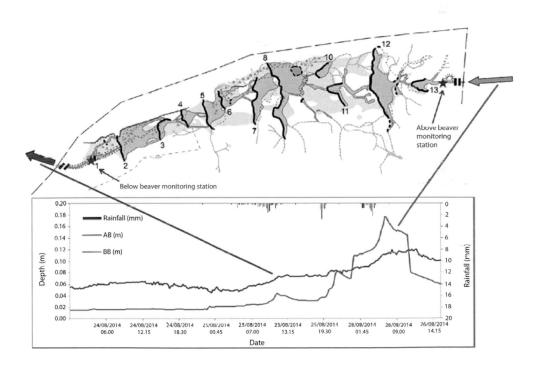

Figure 5.6 Changes in water flow through the beaver ponds at the enclosed beaver trial site, Devon Wildlife Trust.

prevent additional water storage. One fundamentally important aspect of beaver dams is that they are constantly leaking, releasing water into the headwaters, and freeing up flood storage capacity. The structural complexity of the channels created, with multiple dams and fallen trees, creates surface friction which increases lag time and reduces peak flows.

Preliminary data on how the site operates when under saturated antecedent (pre-storm) conditions have demonstrated that when the soil water levels are high and the ponds are also more or less full, a flood amelioration effect is still apparent due to the 'surface roughness' and friction of the site. This stretch of watercourse is now so complex, and full of dams and fallen trees, that however saturated it is, the flood flows though it are still impeded compared with a more defined or engineered channel. This effect is clearly more marked when the site is drier, as there is even more capacity, but the effect is still there in saturated conditions. If such processes were extrapolated throughout a catchment, the potential impacts on flood flow and equally the retention of water in periods of drought could be considerable (Gow and Elliott 2014). The use of Eurasian beavers as a tool to assist the implementation of the EU Water Framework Directive to achieve 'good ecological and chemical status' has been discussed (Törnblom *et al.* 2011; Gow and Elliott 2014) and warrants further scientific investigation, particularly in a British context.

5.4   Beavers in landscape-restoration projects

It is often stated that no other species, apart from humans, has the ability to modify freshwater environments to the same extent as the beaver. Beavers have certainly been doing so for significantly longer than humans, with demonstrated abilities to increase the complexity and biodiversity of freshwater ecosystems (Rosell *et al.* 2005; Stringer and Gaywood in press). This can make the beaver an important factor in landscape-restoration projects directly through their ability to regenerate rivers, create new wetland habitats and improve riparian forest health, but also indirectly by acting as a charismatic flagship species for conservation projects.

Land-managers in Britain have already reported beaver activities as producing positive resource-saving impacts, including reed-cutting, natural tree-regeneration thinning, the creation of open water and willow scrub removal. All these activities are common management practices in wetland landscapes to maintain favourable habitat types for biodiversity.

Bridge Creek restoration project, Oregon

The Bridge Creek project in western Oregon is the location of an innovative habitat-restoration project which has sought to harness the dam-building ability of beavers to conserve endangered steelhead salmon (*Oncorhynchus mykiss*). Historic overgrazing of the surrounding environment and disconnection of the river from its former flood plain has resulted in a significant loss of biodiversity. Bridge Creek had become a single, incised channel, scoured down to bedrock and prone to increasingly significant flood events. A partnership approach to encourage a small, extant beaver population to restore the geomorphic, hydrological and ecological functions of the degraded system by helping the beavers to construct longer-lived dams was developed as a potential solution. Most beaver dams constructed within the incision trench were destroyed during high-discharge events by the full force of floodwater. A simple methodology to assist the beavers to develop more durable dam structures involved installing round wooden fence posts with interwoven wicker in the channel. These structures were designed to function as the crest elevation of active beaver dams. After initial experimentation with this approach in 2008 with 16 structures, a full restoration experiment was implemented in 2009 with 84 structures installed. Within months of their installation, roughly 25% of these were occupied and modified by beavers. As compared to unreinforced dams, they generally fared better through the first major floods, with most either remaining intact or experiencing only minor breaches that were easily repaired by the beavers. Owing partly to the high sediment loads in Bridge Creek, the geomorphic response has been rapid, with net aggradation documented in all reaches, and some degree of flood plain reconnection taking place. Steelhead salmon have returned in increasing numbers to inhabit the dynamic, healthy riverine environment which is being developed by increasing numbers of beavers.

The overall value of the benefits that beavers bring is complex and challenging to calculate. While it is easy to focus simply on the actual costs of beaver mitigation or management work, land-managers and decision-makers involved with such conflicts should be encouraged to develop a more holistic approach which considers the wider costs and benefits of beavers on a catchment, landscape or regional scale.

5.5 Beavers and managed land use

As well as understanding the benefits that beavers can bring, it is also important to accept that their presence can sometimes conflict with human interests and impose a cost in terms of resources (including time and financial), especially in intensively managed landscapes. The idea that beavers are a species exclusive to natural or 'wild' landscapes is fundamentally misplaced. As with many other species, this association was simply a function of such places being remote from human predation. The experience of the authors throughout much of Europe and North America is that beavers readily adapt to highly developed urban and agricultural landscapes where watercourses and riparian vegetation are available.

At low densities, beavers have the ability to blend unobtrusively into an environment, with any conflicts that do arise tending to be localised. During the initial phase of colonisation, they select the most favourable sites, typically larger rivers and lakes, where dam-building activity is rare. As beaver populations grow and their densities increase, successive generations are forced to occupy suboptimal habitats in minor watercourses or anthropogenic environments. In such locations, their presence can become more obvious as environments are modified through a process of dam creation, culvert blockage or other modifications that increase water levels for protection of natal lodges and access to feeding sites. It is generally at this point that conflicts with human land-use interests become more likely.

Figure 5.7 Collapsed beaver burrow in a crop field near to a body of freshwater, Bavaria. (G. Schwab)

Figure 5.8 Collapsed burrow in a flood defence bank, River Isla, Scotland. (H. Dickinson)

Figure 5.9 The effect of beaver activity on a landscape scale, Bavaria: the straight line of a beaver dam is visible in the centre of this photo taken in 2011. While this represents newly created wetland with a leaky dam to enable fish passage, such flooding of arable land may be undesirable. (R. Campbell-Palmer)

Figure 5.10 The site shown in Figure 5.9 in 2014. (R. Campbell-Palmer)

The majority of human–beaver conflicts occur within a relatively slim strip of habitat adjacent to freshwater habitats. In Bavaria, over 90% of beaver conflicts occur within 10 m of the water's edge, while 95% occur within 20 m (Schwab *et al.* 1994). Although conflicts further away than this are possible, they are rare and are usually associated with attractive food sources. Dam creation and its attendant landscape alteration is the most common cause of conflict with an associated requirement for reactive management.

In the fertile lower valleys of Bavaria with their highly developed human infrastructures of roads, raised flood walls, fish ponds, recreational water features, gardens, golf courses, urban areas and forestry, the process of historic land drainage has culminated in a landscape that is dominated by canalised watercourses to support systems of highly intensive agricultural production, and one in which the re-establishment of beavers has proved challenging. Such intricate systems of land drains,

Figure 5.11 Beaver dam activity, upland stream, Norway. (F. Rosell)

ditches and canalised watercourses are essential for this system of land-use; and their failure or blockage can cause significant problems. While living with beavers is neither impossible nor unwelcome, it has tended to result in the formation of both regulated

Figure 5.12 Beaver–human conflicts are more likely in this type of land-use in Bavaria, which beavers will utilise for dispersal and sporadic feeding. These systems are more likely to be dammed, especially if there are adjacent food sources. (R. Campbell-Palmer)

Figure 5.13 Multiple damming to create a series of pools in Scotland. (D. Gow)

beaver-management systems and structures, often underpinned by state funding, or less regulated systems of illegal beaver control carried out by local landowners at their own expense.

5.5.1 Agriculture

Beavers will occupy intensively utilised agricultural landscapes. They have been reported feeding on a range of crops including sugar beet, maize, cereals, oilseed rape, peas and carrots. Beavers will also readily consume the bark, stems, fruit and leaves of fruit trees and soft-fruit shrubs, and may dam areas to access crops more easily (Nolet and Rosell 1998; Campbell *et al.* 2012a). Beaver feeding on the stems of rigid, vascular crops can be identifiable through a cutting pattern at a distinctive 45° angle. This can be very similar to feeding signs produced by water voles and brown hares (*Lepus europaeus*) on lighter plant types. Beaver feeding sites on terrestrial, non-woody vegetation tend to be intensively cropped in patches that are several metres in diameter. Obvious haul-out points and foraging trails from the water's edge to these feeding areas are generally visible and are commonly accompanied by the presence of a scattering of cut stems. Other crops may also be consumed where these are in close proximity to a watercourse. Although the crop-feeding pattern of beavers is generally seasonal and limited to areas within 20 m of the water's edge (e.g. Campbell *et al.* 2012a), they can expand their feeding zones through pool creation, the excavation of canals or burrowing. Beaver feeding on commercially valuable trees or crops adjacent to a water body is relatively limited in extent, and its commercial impact is minor in comparison to other more wide-ranging species such as rabbits, wild boar or deer. In Bavaria, direct crop loss through feeding tends to be low and generally accepted by farmers, with seasonal use of temporary electric fences on small sections of fields bordering fresh water (Schwab and Schmidauer 2003). Beavers do not hibernate. In countries where winter weather conditions can be severe, they rely on gathering piles of woody browse in the autumn which is stored underwater as a food cache to provide them with a winter food resource (Wilsson 1971). The severity of any given winter, coupled with the availability of trees, can dictate the survival rates of beaver populations. This seasonal restriction is unlikely to be as significant a factor for the species in Britain, where the generally milder climate affords the opportunity for constant foraging throughout the winter months.

Figure 5.14 Seasonal feeding on crops. (R. Campbell and R. Campbell-Palmer)

By far the most significant impact of beavers on agriculture is the damming of drainage ditches and/or nearby water bodies so that the backup of water directly floods agricultural land. Additionally, the increased groundwater levels can impede drainage function and cause waterlogging of crops (Schwab and Schmidbauer 2003). These activities cause a more significant economic impact and resultant conflict. In arable landscapes where suitable buffer zones exist, beaver conflicts are reduced (Schwab 2014). Beavers can be dissuaded from feeding in arable crops by a range of management measures (see Appendix C).

In many modern European landscapes, livestock grazing is not as intensively practised as it is in Britain. Commonly, dairy and beef finishing units are permanently housed in indoor units where forage is brought to them daily, while more extensive beef suckler herds or sheep flocks are maintained on higher land away from watercourses. Beavers will re-crop specific areas of grass on livestock pasture to form distinctive beaver lawns. On occasion, beavers have been observed grazing in the company of dairy cattle, which have paid them little attention. There have been a few anecdotal reports from Bavaria of calves falling into beaver burrows, and some injuries have been recorded where hooves have broken through into burrows. Overall, there is very little information from Europe on the collapse of beaver burrows in livestock or equestrian pastures. Livestock may provide a source of infection to beavers for diseases such as cryptosporidiosis. In a large semi-captive enclosure in Devon, beavers have been observed feeding among grazing sheep in the early evening. Where sheep creep feeders or game feeders are positioned within 20 m of the watercourse, beavers will readily enter these to consume the pellets

Figure 5.15 Beaver forage trail to sheep feeder on farm, Devon. (D. Gow)

Figure 5.16 Beaver-cut corn stalk. (D. Gow)

Damming in drainage ditches, Perthshire, Scotland

Over a 12 month period, a farm located in a lowland arable area of Perthshire had over 30 dams built across the site, which included over 13 km of actively managed burns and drainage ditches. The ditches and burns were maintained to support subsurface field drainage of the site. Dams left *in situ* had the potential to raise water levels and impede field drainage as a result. Dams were regularly removed by the landowner but were repeatedly rebuilt in several locations. Weekly monitoring of all burns and drainage ditches on the site is now being undertaken, and dams are removed immediately.

they contain. Where livestock grazing is the predominant land-use, beavers will graze in pastures close to water bodies. Although the flooding of pasture will potentially impact upon its grazing value, the significance of this issue will be dependent on the topography of the surrounding landscape, its land class and the extent of any available buffer zones (these are typically an area of wet woodland or riparian vegetation strip, ~20 m in breadth bordering a water body). While it is possible that the localised felling of trees adjacent to a watercourse could damage stock fencing, there have been few reports of this to date.

5.5.2 Horticulture

Beavers will readily consume the bark, stems, fruit and leaves of fruit trees or soft-fruit shrubs including blackberries (*Rubus fruticosus*), raspberries (*Rubus idaeus*) and blackcurrants (*Ribes nigrum*), and will dam areas to access such crops (Nolet and Rosell 1998). Particularly sensitive species and/or rare specimens should be protected (see section 7.3) as soon as beavers are suspected in an area, since even light feeding may cause significant damage in some species.

Hedges as woody features can provide beavers with a range of feeding, building and residential opportunities. Many hedgerows have seasonally damp or permanently wet watercourses running along their length. Quite commonly, as a result of abandonment, these watercourses can be deeply incised with water flowing several metres beneath the normal ground level. In these deep but narrow environments beavers can create dams of several meters in height which are capable of withstanding significant flows. While most open ditches can be dammed by beavers, the impoundment of shallower ditches will result in flooding which is readily visible, while in deeper gully-type structures the presence of dams may not be obvious for some time. Hedgerows are often planted on soil banks which are exploited by burrowing species such as badgers, rabbits and a range of smaller mammals. In the Parc naturel régional d'Armorique, Brittany, beavers have commonly exploited hedge-banks as burrow locations (Coles 2006).

5.5.3 Woodland and forestry

Beavers can exist in both commercial hardwood and softwood plantations where deciduous tree species, shrubs or vascular plants are available (Pinto *et al*. 2009). While Eurasian beavers do not commonly feed on evergreen tree species, they will occasionally ring-bark conifers or feed on their saplings and low-hanging branches in the late winter/

Figure 5.17 High dam in a deep drainage ditch. (Photo: D. Gow)

early spring. This has been seen particularly in early spring at the Scottish Beaver Trial. The Scottish experience to date demonstrates that impact on woodland will depend on tree species composition, with greater impact on preferred species in riparian areas which, if combined with grazing pressure from other species such as deer, may lead to a shift in tree species composition (SNH 2015). Beaver–ungulate interactions (Hood and Bayley 2009), specifically grazing by deer and the regeneration ability of beaver-felled trees, may require further investigation and management in a British context.

There have been no reports across Europe of nationally significant economic or ecological damage to woodlands caused by beavers (Reynolds 2000); however, this should not detract from the significance of losses to individual stakeholders. In a questionnaire survey of beaver managers and researchers across Europe, Campbell *et al.*

Beaver removal from forest drainage systems, Latvia

In Latvia, the policy of clearing forestry drains of large numbers of beavers has resulted in the development of complete culling strategies to remove all beavers on a substantial scale. This technique begins with the systematic breaching of dams in the lower reaches of drainage systems. The beavers which inhabit these are then flushed from their lodges or burrows with hounds, prior to being shot. Any timber felled across the water course is removed. The riparian woodland at the side of the ditch is felled to a depth of greater than 20 m, and the ditch is then mechanically reprofiled. This process is subsequently repeated until the whole drain network has been cleared (Gackis and Bērziņa 2012).

(2007) determined that most beaver conflicts within forestry plantations were localised and small scale, involving the loss of trees due to flooding as a result of beaver dams, as opposed to significant felling. In the later 20th century, vast commercial conifer forests were planted on recently reclaimed wetlands in Poland and the Baltic states. These landscapes were flat and were drained through extensive ditch systems, with narrow and extensive branches, which were often hand-dug. Beavers were either an extinct or uncommon species at that time, and their subsequent recovery now presents a problem. Localised issues arise where beavers block major culverts, dam drainage ditches or burrow under service roads. Beaver dams, their blockage of culverts and tree-felling across drainage systems all slow the flow of water from the surrounding forest while raising the groundwater table. This activity can flood timber crops and hinder access for workers (e.g. North American beavers in Finnish commercial forests, Härkönen 1999). Monitoring and clearing drains of the resultant debris may require considerable effort. The impoundment of minor ditch systems by beavers is usually inconsequential, although felled trees can block forest tracks, while the creation of standing water bodies which submerge tree roots may result in increased areas of dead standing timber – which, though ecologically important, may cause economic loss of timber crops.

Within their established range, North American beavers fell and consume the bark and foliage of western hemlock (*Tsuga heterophylla*) (O. Rackham, personal communication). It is not known if Eurasian beavers will exploit this commercially cultivated species. Conifers can be felled for construction purposes, but this tends to occur only if few broadleaf trees are available. Where conifer habitats form a significant part of their territorial range, individual beavers may forage quite widely on the woodland floor searching for seedlings, shrubs or vascular plants. While the overall direct impact of beaver browsing in conifer plantations is likely to be insignificant, the effects of beaver damming may present more of an issue (see below). While some evidence of beavers feeding on conifer saplings and side branches near the water's edge has been recorded at the Scottish Beaver Trial, this activity has been seasonal (late winter/early spring) and on a small scale.

The process of modern commercial afforestation began in Britain after the First World War with the establishment of the Forestry Commission to create a strategic timber reserve that would be capable of future economic exploitation. Early planting

Figure 5.18 Damming along a forestry service road. (S. Lisle)

Beaver impacts on productive forest, Norway

In 1993, Parker *et al.* (2001a) investigated the proportion of the landscape and productive forest affected by beaver in a typical montane landscape in southeastern Norway, with forest vegetation similar to that originally found in much of Scotland. The study area (34.7 km²) consisted of 648 managed forest stands. Immigrating beaver first became established on the study area in 1965, and the population peaked around about 1990. Nineteen beaver territories (0.55/ km²) were occupied in the study area in 1993, and 19 additional sites had recently been abandoned. Only 0.1% of the productive forest in the study area had been inundated from beaver dams. The low proportion of impounded area was due mainly to the beavers' use of naturally existing ponds and lakes as sites for lodge-building, plus the mountainous nature of the landscape that led to small impoundments where dams were constructed. No Norway spruce (*Picea abies*) or Scots pine (*Pinus sylvestris*) was found felled. Beavers had removed approximately 50% of the birch (*Betula pubescens*) and 60% of the Eurasian aspen (*Populus tremula*) ≥10cm diameter breast-high (the minimum tree size defined in the study as having commercial value) from that proportion (2.6%) of the total study area where beaver felling occurred. This amounted to approximately 1.3% of the harvestable birch and aspen in the total study area. Since heavy thinning of most young birch and aspen to enhance conifer growth in productive forest was part of the forest-management plan, felling of young birch and aspen by beaver in many stands likely constituted production benefits for the forest-owner, rather than damage. In addition, the sale of beaver-hunting helped to reimburse some of the income loss incurred from tree-felling and dam-building. However, Finnish foresters have expressed greater concerns over beaver damage than Norwegians (Härkönen 1999). This is most likely due to smaller mean forest property size, and therefore damage experience is relatively higher per individual forester (Parker *et al.* 2001a).

policies were largely based on non-indigenous, fast-growing conifer species such as Sitka spruce (*Picea sitchensis*), Norway spruce (*Picea abies*) and Japanese or hybrid larch (*Larix* spp.). Many of these plantations were situated on poor or relatively marginal land which often required extensive drainage. These drainage systems may now be fringed by indigenous species of riparian broadleaf trees, which provide exploitable habitats for beavers. Although their dams can be drained or removed for essential management purposes, consideration should be given to retaining them and allowing these areas to develop into ecologically rich, variable habitats, which may have multiple benefits in a forest environment.

Beaver dams have the capacity to flood considerable areas of flat land (e.g. Zurowski and Kasperczyk 1986). Where watercourses are steeper in gradient with higher banks in narrow valleys, the capacity for beaver activity to alter or create habitats on a significant scale is more limited (Parker and Rønning 2007). In managed woodlands which contain extensive wet-woodland communities requiring intervention for ecological reasons (e.g. Devon Beaver Project), the presence of beavers could be economically beneficial. In woodlands where nature conservation or recreation forms a primary function, the habitat changes brought by beaver activity could also be desirable.

Figure 5.19 Beaver feeding station with beaver-cut conifer, lily stalks and finer-peeled sticks, Knapdale Forest, Mid-Argyll. (Scottish Beaver Trial)

Figure 5.20 Camera trap image of beaver dragging freshly cut conifer. (Scottish Beaver Trial)

The composition and character of riparian woodland is variably impacted by beavers. While some tree species such as oak (*Quercus robur*) are relatively tolerant of the partial or seasonal inundation of their root systems, others such as elm (*Ulmus* spp.) or juniper (*Juniperus* spp.) cannot tolerate extended periods of inundation (Niinemetus and Valladares 2006). Although alder (*Alnus* spp.) and willows (*Salix* spp.) are both adapted for survival in wetland environments, the former cannot survive for long in pools or ponds, while the latter is more tolerant of flooding (Niinemetus and Valladares 2006). Even so, few European tree species survive permanent flooding, and dead standing trees will result. Similarly, the character of the tree and shrub plant community will alter over time where regular beaver browsing is sustained. Strongly favoured tree species such as aspen, willows and rowan (*Sorbus aucuparia*) (Jenkins 1975; Haarberg and Rosell 2006)

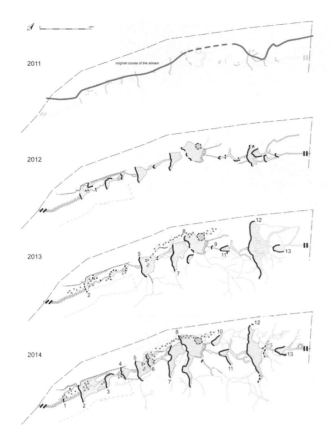

Figure 5.21 Damming progression at an enclosed trial site, Devon Wildlife Trust. Solid black lines represent beaver-constructed dams with the resultant open water shaded in dark grey. The original, single water course running through this site can be clearly seen marked in black in 2011. Copyright South West Archaeology. Inset: Aerial view of the site. (Mark Elliott, Devon Wildlife Trust)

will be reduced in canopy height and changed in structure in favoured feeding locations (Johnston and Naiman 1990).

Beavers will commonly re-coppice trees or shrubs that they have previously browsed. This activity can be seen in the tree morphology and give some indication of the historical activity of beavers at a specific location. Some tree species, such as aspen, respond to beaver felling by suckering, sending out large numbers of suckers from their root systems which rapidly form dense thickets of young shoots when a main stem is felled (Jones *et al.* 2009).

Although poplar (*Populus* spp.) is no longer commercially valuable, in Bavaria ring-barking by beavers of many large trees has proven to be socially unacceptable due to the 'untidy' appearance of the resulting environment. In Britain, the cultivation of cricket-bat willow (*Salix alba* 'Caerulea') in the flat riparian landscapes of eastern England could be affected by beavers if they established in this region. Plantations next to watercourses might require mesh tree guards or beaver-exclusion fencing (Appendix C).

Beavers are known to exploit a very broad range of deciduous trees and shrubs. On the River Elez in Brittany, where a beaver population was established in the mid-1960s

Figure 5.22 Beaver-coppiced birch. (D. Halley)

Figure 5.23 Willow regenerating from a beaver-felled trunk. (D. Gow)

and is currently estimated to number around 120 individuals, virtually no visible feeding activity is evident on the abundant oak woodland present along the river banks. In higher-lying stream systems, as seen on the River Elez, more unpalatable tree species (e.g. oak and hawthorn *Crataegus monogyna*), commonly remain undisturbed by beaver feeding activity, although they can be affected when their root systems are submerged, and when these trees are felled for construction activities. On occasion, some less desirable food trees will show the signs of a few exploratory bites removing their bark.

5.5.4 Fisheries

There is a lack of published data on interactions between beavers and fish in Europe, with most studies originating from North America. The potential benefits of beavers and their associated activities have been discussed in a number of publications (e.g. Macdonald *et al*. 1995; Collen and Gibson 2000; Rosell *et al*. 2005; Kemp *et al*. 2010; Kemp *et al*. 2012). Their potential negative impacts have also been reviewed and include dam creation activity acting as a barrier to fish movement, physical damage and silt retention which can eventually cover valuable habitats (Knudsen 1962), and fish mortality due to a reduction in oxygen levels in beaver ponds (Kemp *et al.* 2010). The capacity of beaver dam systems to interfere with migratory fish movements, particularly salmonids, is a contentious issue receiving particular interest in connection with beaver reintroduction to Scotland. The effects of beaver dams on fish migration are likely to depend greatly on both the topography and hydrology of a given site. Other effects of damming, such as changes in the water velocity or temperature, have also been considered in terms of their potential influence on fish spawning and population growth (Collen and Gibson 2000; Halley and Lamberg 2001; Mitchell and Cunjak 2007; Malison *et al.* 2014).

In Bavaria, beaver activity in wintering fish ponds can in rare cases keep the fish moving to the extent that they lose weight and consume more oxygen. In combination this may lead to fish-stock loss (Schmidbauer 1996; Schwab and Schmidbauer 2003). It should be noted that many disturbances can cause this effect, including the activities of predators such as the North American mink. It would require further scientific investigation to determine if this is a significant concern associated with beavers. It has been suggested that beaver activities where water flow is slowed down could lead to the dissipation of dissolved oxygen (McRae and Edwards 1994), with a resultant reduction in the potential stocking capacity of a fishery.

The freshwater ecosystems of Northern Europe are adapted to the presence of beavers. Where beaver populations re-establish, their presence throughout a river catchment and their behavioural activities can result in a complex cascade of both advantageous and disadvantageous opportunities for a wider range of species. The presence of beaver dams should be viewed as a dynamic process of construction and abandonment, which over time results in a continual adjustment between fish migration and water-body discharge (Taylor *et al.* 2010). Research has shown that this increase in lentic (still water) habitats can benefit both juvenile trout, which have a strong affinity with pool environments, and large migrating adult salmonid species, where ponds can offer an important refuge (Collen and Gibson 2000). Beaver dams do not present a solid barrier to fish such as salmon if there is sufficient water passage over or around the dam itself for a fish to swim up and/or a deep enough column of water to enable fish to jump (Mitchell and Cunjak 2007; Thorstad *et al.* 2008; Malison *et al.* 2014). Beaver dams are 'leaky' structures with changeable flows. Water passes around, over, through and under dam structures (Lokteff *et al.* 2013). The majority of published research, though often citing beaver dams as barriers to fish movement, does not indicate that beaver dams are routinely impassable to anadromous species (species that spawn in fresh water but spend most of their life at sea) (Kemp *et al.* 2012). In Norway, beavers commonly share the same watersheds with salmon and trout fisheries. Populations of hatchling salmon have been recorded above a series of beaver dams up to 1.6 m high over a 600 m stretch of stream, indicating the passage of adult fish (Halley and Lamberg 2001). There is very little perception of any beaver–salmonid issue in Norway (D. Halley, personal communication 2014), and so almost no research has been undertaken despite the economic importance of salmonids. Beaver populations are present at carrying capacity on many salmon rivers (Kemp *et al.* 2010).

Other studies demonstrate that juvenile salmonids are able to pass over or through dams to continue their downstream migration (Bryant 1984; Swanston 1991; Alexander 1998) and that adults use side channels created by the diverted flow to bypass dams. Lokteff *et al.* (2013) found that Bonneville cutthroat trout (*Oncorhynchus clarkii utah*) and brook trout (*Salvelinus fontinalis*) readily bypassed even large beaver dams, and argue that many concerns over fish passage are largely speculative. Parker and Rønning (2007) concluded that, due to the low frequency, small size and short lifespan of beaver dams on spawning tributaries in Norway, their effect on salmonids would be negligible even assuming any impediment to movement (which was not tested), and that this feature of their presence should have little overall impact on the reproductive success of salmonid populations in the long term. This conclusion is supported by field studies of Pacific and Atlantic salmon populations in North America (Collen and Gibson 2000). For example, fish–beaver interactions have been actively managed in watercourses in parts of California where beavers were removed to encourage population growth of the endangered Paiute cutthroat trout (*Oncorhynchus clarki seleniris*: Snyder 1933) (Hunter 1976). The role that any particular system of beaver dams may have on fish movements will depend on the character of formation, its size and location, as well as the river flow and seasonality (Halley and Lamberg 2001; Rosell *et al.* 2005; Kemp *et al.* 2010; Malison *et al.* 2014).

In Britain, much concern regarding the potential deleterious impacts of beavers has focused on the ability of their dams to impact negatively upon migratory fish. As noted by Taylor *et al.* (2010), the issue of beaver dams preventing any movement of migratory fish is ultimately related to the historical degradation of watercourses. Beaver dams stabilise water levels and slow drainage rates in systems (Gurnell 1998; Collen and Gibson 2000) which have lost their natural resistance to variation in rainfall (Bobba *et al.* 1999; Walsh and Kilsby 2007). This process, beneficial to salmon, may be critical in the future if climate change results in increasing water shortages.

Beavers can be advantageous for the following reasons:

- The deeper water of beaver ponds can provide important juvenile rearing habitats (Scruton *et al.* 1998), as well as important habitat for adults during the winter (Cunjak 1996). During droughts, beaver dams can provide important refugia for fish (Duncan 1984).
- The presence of dams reduces siltation of spawning gravels below the impoundment (Macdonald *et al.* 1995).
- Beaver building and foraging activities result in more complex underwater environments that can benefit fish (Kemp *et al.* 2011). Submerged habitats that result from the presence of large amounts of woody debris can be used by fish as refugia from agile predators such as cormorants (*Phalacrocoracidae* spp.), mergansers (*Mergus* spp.) or common otters (Rosell *et al.* 2005).
- Decaying of felled woody material provides feeding sites that are rich in aquatic invertebrates (Hägglund and Sjöberg 1999). These more complex microhabitats also provide good spawning grounds for a number of fish species (Collen and Gibson 2000). Greater food availability has been shown to increase growth rates in fish (Murphy *et al.* 1989; Sigourney *et al.* 2006).

The reintroduction of beavers has the potential to result in a number of conflicts with native fisheries, including:

- Beaver dams have been documented to affect fish migration, particularly during periods of low water flow. In Lithuania, rapid beaver population growth and resultant dams have resulted in increased disturbance to fish migration (Ratkus 2006); and in North America some dams have been found to limit temporarily adult and juvenile fish movement (Alexander 1998; Cunjak and Therrien 1998). Concerns over the ability of salmonids during spawning periods and other species, such as large adult lamprey, to pass beaver dams have been raised.

- Reproductive disruption can be caused to anadromous salmonids. The formation of dams reduces stream velocity and increases silting upstream of the dams, which can impact on available spawning habitat (Knudsen 1962) and alter the distribution of riffles and runs which are the preferred habitat of salmonids (Mitchell and Cunjak 2007). Selective felling may also reduce the species composition of leaf litter entering streams, which could alter their insect fauna and, consequently, the nutritional base for juvenile fish (Collen and Gibson 2000).
- The reduction of bank-side vegetation and its associated shade can result in increasing water temperatures, which could be damaging to salmonid populations. However, such an effect is particularly site dependent, as increased light can also increase in-stream photosynthetic production, hence increasing invertebrate and fish biomass.
- Increased temperatures found above and below the impoundments may be suboptimal for salmonids and are likely to reduce dissolved oxygen levels (Avery 1983), which could be of particular relevance in the light of climate change. Alternatively, beaver dams provide deeper water, which stays cooler at the bottoms of ponds during summer months.

Conflicts associated with dams are not always permanent. For example, the removal of a dam will result in the displacement of lighter silts while heavier materials remain in place, creating riffles and potentially spawning beds. Field research also indicates that, on most watersheds, most beaver families do not build dams. For example, in 2003 on the Numedalslågen River system in Norway, beaver dams were only constructed in 5 out of 15 tributary stream territories (each on one of a total of 51 tributaries navigable by anadromous salmonids); there were no dams on the main river, where most beaver territories were located (Parker and Rønning 2007). This is because beavers strongly prefer slow-flowing, wide river habitats which do not require a significant expenditure of energy on dam construction (Halley and Rosell 2002). Beavers are highly territorial. They react aggressively towards members of unrelated families and this requires family units to maintain a territorial area (Piechocki 1977; Tinnesand *et al.* 2013) which completely provides for their seasonal feeding requirements. Their densities are limited by the availability of these habitats – and, as a result, few dams are normally functional at any one time (Cook 1940). A wide body of research indicates that the view that beaver dams are routinely impassable to anadromous species is not sustainable. Techniques for managing water levels associated with dams are discussed in the following section; however, more research has been called for to assess whether fish are able and willing to pass through structures such as flow devices.

5.5.5 Engineered environments

Beavers can readily adapt to highly developed urban landscapes where woody browse and other riparian vegetation is plentiful. While the initial arrival of beavers often invites much media attention and is commonly welcomed by wider society (Müller-Schwarze 2011), their activities over time can be negatively perceived when, for example, they consume garden plants and ornamental or fruit trees (Campbell *et al.* 2007). On larger rivers that flow through urban environments, their presence in areas of well-established riparian woodland or islands is commonly unobtrusive. Beavers demonstrate flexibility in their feeding habitats and can survive in a broad range of fresh (e.g. the inner city of Bratislava: Pachinger and Hulik 1999) and even brackish water areas. They are resilient to changing conditions and can create multichambered lodges to provide for annual patterns of water fluctuation that extend well up a bankside. They can repair damaged dams swiftly (Wilsson 1971).

Large artificial dams, especially hydroelectric dams, can form significant barriers to beaver population spread, given their reluctance to travel larger distances away from water. However, there have been some reports of beavers travelling impressive distances. Most likely these were dispersing subadults searching for suitable territories away from areas of higher beaver densities, travelling over land to navigate around such structures. To address this and encourage spread, and reduce population fragmentation, management authorities in France have installed 'beaver ladders' at several hydroelectric dams (Office Nationale de la Chasse 1997).

Figure 5.24 Beavers can utilise artificial systems; this beaver pair were removed after attempting to build a lodge in a water drainage pipe running under a sewage plant. (Günter von Lossow)

5.5.5.1 Flood banks

Few, if any, of Europe's larger river systems remain in an unmodified state. Human engineering over time has commonly confined most of their lengths between flood walls. This system of river engineering was developed at a time when beavers were largely extinct. Where embankments are set back from the main channel by a distance of greater than 30 m, the impact of burrowing animals is insignificant. Where flood banks are immediately adjacent to the main water channel, the excavations of semi-aquatic rodents, including beavers and water voles, and where found in mainland Europe, non-native species such as coypu (*Myocastor coypus*) and muskrat (*Ondatra zibethicus*) can be potentially damaging in friable banks. Likewise on their landward side the activities of terrestrial species such as badgers, foxes and rabbits can produce a similar result. Beaver burrows tend to be large and can end in sizeable chambers. Although the route of these structures is occasionally visible, the position of many others is difficult to determine.

Figure 5.25 Digging by beavers leading to areas of exposed flood-bank protection measures against a range of burrowing mammals, Bavaria. (R. Campbell-Palmer)

Figure 5.26 Flood-bank blow-out encouraged by beaver burrowing, River Isla, Scotland. (R. Campbell-Palmer)

5.5.5.2 Sewage works

Settlement beds for sewage in close proximity to a watercourse can be accessed by beavers. These habitats contain a reliable water supply and are commonly surrounded by lush vegetation. There is no indication that sewage odour is a deterrent for beavers. The potential blockage of open water supply for these facilities by beavers may be a problem which requires management intervention. If water supplies cannot be protected from blockage by grilles or flow devices (see section 6.1.4), then the most durable solution is the installation of beaver-exclusion fencing (see section 6.3.4). In situations where this is not suitable, capture and removal will be required. This may be an ongoing process, with any captured beavers being translocated or culled as appropriate (see section 6.4.4).

Figure 5.27 Newly built sewage pool with planned natural reed-bed filtration (far right); sunken weldmesh (middle) to protect banks from any burrowing activity from nearby beaver family (lodge visible in distant left of photo). (R. Campbell-Palmer)

Figure 5.28 Standard fencing surrounding sewage plant, largely to exclude humans. Beavers have pushed underneath the bottom of fencing repeatedly to form a forage trail to feed. (R. Campbell-Palmer)

5.5.5.3   Canals

In recent years, there has been a rapid growth in the restoration and use of canals for recreation. Beavers readily use large, canalised water bodies in mainland Europe. Where the banksides of these canals are reinforced with rocks, sheet metal, concrete or heavy wooden piling, there are few conflicts with beaver activity unless there are outflows or weirs that can be dammed (DVWK 1997). Beavers can create lodges on artificially reinforced banksides where building materials are available and will readily feed on adjacent vegetation. In the Netherlands, during the early 2000s when the reintroduced beaver population was still small, nature conservation bodies removed sections of hard reinforcements on some canals to create shallow bays to allow beavers to access adjacent woodland (V. Dijkstra, personal communication 2014).

It is likely that, where canals contain bay-type features set back into the surrounding landscape with riparian vegetation and soil banks, these would be attractive habitats for beavers. If these are additionally connected to wider wetland, river, ditch or other riparian networks, then the immigration of beavers into a canal system will be inevitable. The principal concern regarding the existence of beavers in canals in Britain would be their ability to create burrows that may cause structural damage to retaining banks. Although mitigation is possible, it tends to be expensive to install, especially given the length of bankside that would need to be reinforced. As beaver populations increase, their regular removal from these environments or their exclusion could become necessary.

5.5.5.4   Roads/culverts/drains

Although beavers are commonly killed on roads, some beaver families can live in close proximity to major road systems for many years without regular casualties or disturbance. While most dispersing beavers readily follow watercourses, they will sporadically travel overland in the early spring. It is possible that some of the fencing designs which have been developed to mitigate effectively against otter road casualties (e.g. fencing to prevent access to roads, under road passages, weir ladders: Chanin 2006) could be adapted for beavers as appropriate.

Through dam-building, beavers can threaten different types of upstream properties. One of the most common and serious problem tends to be clogged road culverts. With the road acting as a pre-existing dam, water can quickly accumulate behind this barrier. This can impact the road structure, causing damage to property upstream and affecting land-use there (Jensen *et al.* 1999; Boyles and Savitzky 2009). Culverts are a common feature in developed landscapes. Although designed in different styles and of varying dimensions, they provide a basic drainage function for watercourses crossing beneath tracks, roads or buildings, or through settlements. Impacts tend to be more significant in flatter and more intensively managed systems. While the simple removal of dams or impoundments can sometimes be effective, beavers can be persistent in their reconstruction. Continual dam- or beaver-removal programmes are expensive and time consuming (Jensen *et al.* 1999), particularly in areas of suitable beaver habitat which are easily recolonisable (Houston *et al.* 1995). As beaver populations grow and stabilise, the simple removal of dams (without the implementation of suitable flow devices) to protect culverts is rarely effective or economic unless it is accompanied by a policy to reduce population numbers.

Where streams flow under, or run alongside, minor roads there is the potential for beaver-felled trees to cause problems. Potential solutions to this issue over limited areas

Figure 5.29 Blocked culvert, with higher, retained water levels seen in the background (left). The beaver dam has been built in front of the culvert as opposed to within it, so water still flows through it. The land lies in a flood plain and is not considered valuable enough by owners to employ any management. (R. Campbell-Palmer)

Figure 5.30 Beaver burrow undermining a road in North America. Here, anti-burrowing fencing has subsequently been erected to prevent further burrowing. This burrow was later sealed to prevent any road collapse. (S. Lisle)

would be to coppice, protect or remove any likely food-tree species within felling distance of the road. In areas where water, gas or sewage pipes, and electricity or telephone cables are installed next to fresh water, these can be damaged by gnawing and may require protective measures such as sunken or exclusion fencing, or beaver removal. There have been a few examples in Bavaria of beaver burrows resulting in the limited collapse of 3–4 m sections of minor roads where these were immediately adjacent to a stream bend.

5.5.6 Recreational areas and water bodies

Beavers will readily occupy environments that are regularly used for recreational activities such as swimming, leisure boating, sunbathing, jet-skiing and canoeing, emerging in the quieter hours to forage. Streams and ponds in places such as golf courses or parks are also viable habitats. There are very few records of beavers attacking dogs or people. In March 2013 in Ostromechevo, Belarus, a man bled to death after cornering a beaver whose bite severed a major artery in his leg. He is the only person known to have died from a beaver bite (theguardian.com 29 May 2013).

In Bavaria, where old gravel pits are commonly utilised for recreational pursuits, there are no examples to date of accidents resulting from watersports participants colliding with beaver-felled timber. In some sites where the felling of trees could impact upon access roads or people, a proactive coppicing regime has been developed for small, individual trees. Larger trees tend to be protected with expandable wire mesh tubes as a precaution (see Appendix C.5).

Occasionally, beavers may occupy ornamental ponds. While in most cases the duration of their presence is limited, they can radically alter the aesthetic appearance of these features by felling ornamental trees, burrowing or feeding on garden plants. Unsurprisingly, these activities cause significant annoyance and upset. Ornamental gardens and arboreta are relatively common features in British landscapes, with some being of international importance. Those which connect to freshwater courses could have issues with beaver activity should populations increase. While large specimen trees in the vicinity of watercourses can be readily protected, this may be harder to effect for multi-stemmed shrubs or other palatable vegetation. Although knowledge of beavers' preferred food species in Europe is well established, it is likely they will utilise others with which they are generally unfamiliar. For example, a pair of beavers at the Escot estate in Devon selected bay (*Laurus nobilis*), a Mediterranean species, as a food resource immediately following their release into an enclosure of several acres where willow and aspen were readily available (J.M. Kennaway, personal communication 2014).

Key concepts

- Beavers create dynamic mosaics of habitats with a range of ecological and economic benefits, but this can lead to conflicts, especially in modified land-scapes. Any management should be implemented as early as possible to reduce the extent of the problem and the expense of rectification.
- Scientific evidence demonstrates that beavers have a net positive effect on bio-diversity, though this is variable and some species of conservation value may be negatively affected, with concerns raised over aspen, freshwater mussels and lampreys, for example.

- The flood-alleviation potential of beaver activity should be further investigated in a British context.
- The majority of human–beaver conflicts occur within a relatively slim strip of habitat adjacent to freshwater bodies; therefore the establishment of riparian zones would reduce the likelihood of many conflict situations.
- Flooding and waterlogging are more likely management issues in commercial forestry as opposed to loss through felling.
- Damming is probably the most controversial issue concerning beavers and fisheries, especially if this affects fish migration during the spawning season. Beaver dams are not impenetrable and can be dynamic in nature; therefore any impact is likely to be temporary and highly dependent on water flow.
- Beaver burrowing into artificial structures such as canals, flood banks and roads, though generally uncommon, can cause significant impacts when it does occur. Preventative and reactive management should be implemented as soon as issues are identified.

6. Managing beaver impacts

Techniques for the effective management of beaver impacts are now well developed across Europe and North America (Appendix C). These will have an obvious application in Britain if beaver populations are permitted to survive and expand. Many of these techniques have been developed in response to both legal constraints and a wider social interest in non-lethal wildlife-management solutions. It is, however, worth noting that any mitigation work causing disturbance to this species, direct damage or destruction of its breeding and resting sites, or killing and possession without a licence are likely to be in contravention of the Habitats Regulations (SNH 2015). Additionally legislation may vary depending on the protection status of the area, the impact of the management technique on other protected species (e.g. passage of Atlantic Salmon), and associated licensing implications. Therefore the emphasis is on the reader to seek advice from the relevant authorities in advance. It is likely that the efficacy, cost and legal considerations surrounding the use of any technique will change over time. In Campbell *et al.*'s (2007) survey of European beaver-managers, non-lethal mitigation constituted the majority of management practices. While this may reflect the conservation status of this species, it also offers a more practical solution, as the culling of problem individuals in a highly territorial species merely creates a vacuum to be filled by dispersers (Campbell *et al.* 2007). Continual dam- and/or beaver-removal programmes are expensive and time consuming (Jensen *et al.* 1999), particularly in areas of suitable, accessible beaver habitat (Houston *et al.* 1995). However, this is not to say lethal management will not be required over time as higher population densities of beavers are reached.

Many management methods exploit the general reluctance of beavers to move far from water when foraging. A distance of 150 m between two streams has been shown to have a strong deterrent effect on the chances of beavers colonising from one to another (Halley and Rosell 2002; Halley *et al.* 2013). In Sweden, this effect has been shown at a national scale (Hartman 1995). In France and Switzerland, the authorities have built special 'beaver ladders' that make it possible for beavers to pass large artificial dams and other barriers in heavily regulated rivers, though beavers have also been observed travelling overland to traverse human-made dams in Norway, including large hydroelectric plants. The motivation for some beavers to travel larger distances than generally expected should not be underestimated; however, this motivation will vary between individuals, available resources and reproductive status, i.e. dispersers vs. breeding family members, and with population density, i.e. if there are no free territories, beavers will travel further.

When human–beaver conflicts are analysed in cultural landscapes, it is clear that the majority arise within a relatively slim strip of habitat adjacent to freshwater habitats. In Bavaria, for example, over 90% of beaver conflicts occur within 10 m of the water's edge, while 95% occur within 20 m. Although conflicts further away than this are possible, they are rare and are usually associated with an attractive food source. A study of beaver impact on woodland in a Scottish context over a 4-year period found that most effects on trees were within 10 m of fresh water (Iason *et al.* 2014). Any considered process of long-term beaver management can most effectively be planned for through the establishment of buffer zones around freshwater features. This policy is often already established in many European countries due to wider environmental, economic and social benefits. It is likely that, if this process were to be generally adopted within Britain, then the potential for human–beaver conflicts would dissipate. The planting-up of these areas with native riparian tree and shrub species, or more cheaply, the encouragement of natural regeneration (often absent due to land management practices and grazing pressure), can reinstate suitable beaver habitats, improve riparian ecology and assist in the capture of undesirable sedimentary runoff. This process of habitat-provision in the immediate vicinity of a water body will reduce the requirement for more distant foraging and limit the impact of other activities such as burrowing and canal construction.

This long-term strategy will not always be practicable where flood banks exist, in artificial and/or heavily modified and managed water bodies, or where land is considered too valuable for buffer zones to be created. Once reintroduced into a river system, beavers will, over time and in the absence of any significant physical barriers, spread to occupy any accessible habitats throughout the entire catchment. If this is undesirable, the only practicable solution is to limit beavers to particular areas through a constant, consistent process of removal via trapping or direct culling, to create 'beaver-free zones'. Where densities rise and/or populations expand, this requirement may become more common. The legality of this approach is likely to change over time, so the latest advice should be sought from the relevant statutory bodies before undertaking any process of this type. The implications of each management technique require careful study to ensure they are legal. A licence may be required to undertake certain types of management. Although potentially unpalatable in Britain, hunting or the selling of the right to hunt beavers has been demonstrated in some European countries to be a flexible and cheaper management option, and has previously been an effective method to control populations in Norway and Finland for example (Lahti 1997, Ermala 2001; Parker *et al.* 2002b; Parker and Rosell 2014). However, the validity of such methods in actually maintaining tolerable population levels is questionable (Hartman 1999; Parker and Rosell 2012). In Norway, such an approach affords landowners the power to manage the impacts of both 'problem' beavers and population size with minimal bureaucracy while potentially gaining benefits from culling income derived from having beavers on their land (Parker *et al.*

Table 6.1 General beaver-management techniques according to rapidity of implementation and likely ease of application to land managers. Longer-term techniques are more likely to be subject to licensing procedures and/or wider multiorganisational collaboration.

Immediate management techniques	Longer-term management techniques
Individual tree protection	Trapping (live and lethal)
Deterrent/protective fencing	Translocation
Removal/management of non-lodge-protecting dams	Culling/beaver-free areas
Canal and non-residence burrow management	Flood-bank protection/realignment

2001a; Parker and Rosell 2003). In developed Western European landscapes, the specific targeting of 'problem' individuals appears to be a much more effective conflict resolution than a randomised approach to reducing numbers in a wider population.

6.1   Damming activity and associated management techniques

6.1.1   Ecology

Beavers can form extensive wetland environments in flat landscapes where they are not removed or any damming activity is drained or water levels managed by people. Commonly, dams can be limited in extent by topographic features in a landscape such as narrow incised watercourses. Some of these dams result in no major immediate or even

Figure 6.1 Beaver impoundment in a flat landscape: shallow but extensive dam structure, winter, Bavaria. (R. Campbell-Palmer)

Figure 6.2 Beaver dam during a spring spate, Norway. (D. Halley)

long-term impact for human activities. The establishment of a dam in a narrow location in a flatter or U-shaped valley can result in more extensive upstream flooding. Depending on topography, damming may result in a series of smaller, tiered pools maintained by multiple dams. Further details on behaviour can be found in section 3.6.

It should be noted that, when beavers build dams, there is a dynamic process of natural dam-breaching and restoration as beavers move around the landscape. Though unmaintained dams usually decay within a few years, dams built on low-gradient, small streams or drainage ditches sometimes remain intact for many years, apparently without maintenance, eventually becoming overgrown with vegetation and possibly permanent. Most beaver dams breach at a single point (pinch point) at a narrow outlet which can be easily repaired by the beavers if they so choose. Where they are not repaired, then a watercourse will rapidly form through the silt bed of any former pools. This dam dynamism is also one explanation why fish populations are thought not to experience significant migratory or reproductive issues over the long term.

Figure 6.3 Established beaver dam during spring spate and breach: upstream and down-stream views, Norway. (D. Halley)

6.1.2 Benefits

A variety of factors determine the hydrologic permeability of specific beaver habitats. These include landscape topography, seasonality, and abundance of dry or partially dry beaver ponds, the density of vegetative detritus and the distribution of absorbent plant species such as sphagnum moss (*Sphagnum* spp.). The environments that beavers generally create have a natural capacity to absorb and slow the progress of water during flood events (Elliott and Burgess 2013). Experience with an enclosed beaver release in Devon suggests that there may be seasonal patterns of dam maintenance whereby longer dams with shallow pools are refurbished in the winter months when water is abundant, while shorter, taller structures in the deeper channels become the principal focus of their summer activity (B. Northey, personal observation). This behaviour would naturally complement an effective system of water dissipation in winter and retention in summer.

Damming most obviously creates new wetland habitats for a range of plant and animal species. With the flooding of tree roots resulting in increased areas of dead

Figure 6.4 (Left) Floating pontoon enabling the general public to obtain a unique view of a beaver dam. (Right) Onsite interpretation explaining beaver field signs. (Scottish Beaver Trial)

standing wood, such habitats and associated features may be desirable in some areas. By allowing beavers to dam, these habitats can be created more readily and without the costs of some habitat-restoration projects.

Beaver dams and field signs along forest footpaths can also be perceived as an asset. For example, in Knapdale Forest, Mid-Argyll, a floating pontoon has been created by the Scottish Beaver Trial partners as part of a forest walk around a beaver dam (Figures 6.5 and 6.6). In extensively afforested environments, substantial beaver populations have the capacity to alter the ecological character of landscapes on a significant scale. Although this process can impact upon whole faunal and floral communities (Gurnell *et al.* 2009) through the development of dynamic habitats which vary radically over time, in other situations their presence can be more unobtrusive.

6.1.3 Issues

Along with any benefits, it is important to accept that beaver damming can conflict with human interests and impose a cost in terms of resources (including time and financial impacts), especially in intensively managed landscapes. The potential blockage of water in some areas may be particularly problematic, e.g. where this affects agricultural land drainage or water treatment plants. Beavers can construct new dams quickly, and generally use fresh material each time, so regular monitoring of a sensitive site for any rebuild is often required.

The most significant impact of beavers on agriculture is the potential impact of dams in water bodies relied upon for drainage of agricultural land (see section 5.5.1. for more details of impacts).

Some beaver dams have the potential to impact migratory fish movements. The importance of salmonids to Scotland's fisheries sector results in this being an issue of specific interest in Scotland. This has been reviewed by the Beaver–Salmonid Working Group; their findings were presented in their final report for the National Species Reintroduction Forum (BSWG 2015). Some studies have documented beaver dams affecting fish migration, particularly during periods of low/summer water flow, in Lithuania (Ratkus 2006) and North America (Alexander 1998; Cunjak and Therrien 1998). Changes to water velocity and temperature and siltation above dams, impacting salmonid

Figure 6.5 Beaver damming of a land drainage ditch on flat agricultural land, Kirriemuir, Scotland. On the right the dam can be seen extending into arable field in an attempt to slow overflow from the original (and larger) dam structure within the drainage ditch, behind the fence on the left. (R. Campbell-Palmer)

spawning grounds, have also been raised as potential issues. Such effects of beaver dams on fish migration are likely to depend greatly on both the topography and hydrology of a given site. The location, composition and age of either a set of tiered dams or a single beaver dam will all have an impact (Halley and Lamberg 2001; Rosell *et al.* 2005).

Research on the closely related North American beaver has shown that some dams, particularly in low-gradient systems, can temporarily limit adult and juvenile fish migration, particularly during low water flow (Alexander 1998; Cunjak and Therrien 1998). In Lithuania, the rapid growth of the beaver population has led to an increase in the number and distribution of dams, with a resultant increase in the disturbance of fish migration (Ratkus 2006). Alternatively, beaver dams can improve fish habitat below dams through creating silt-free spawning habitats and areas of deeper water, with cooler temperatures at the bottom during the summer months (Gurnell *et al.* 1998; Rosell *et al.* 2005).

6.1.4 Management options

Any land-manager considering removing a beaver dam or installing a flow device on natural watercourses would be advised to liaise with the relevant regulatory authorities such as (in Scotland) the Scottish Environmental Protection Agency (SEPA) and SNH. SEPA considers that beaver dams can be partially or completely removed from watercourses using hand tools, ropes or grapnels, without prior authorisation by SEPA under the Controlled Activity Regulations (CAR). Such work must be completed without causing pollution. Any dams on artificial drainage channels are not subject to a requirement for prior authorisation by SEPA under CAR (SEPA 2014). Currently the Environment Agency does not have a specific position on beaver damming, but should damming occur, any resultant impact on flooding or fish passage would be assessed and practical course of action sought. If any in-channel or floodplain structure is proposed, then the necessary consent from the Environment Agency or Local Flood Authority would need to be sought.

Considerations put forward by SEPA (2014) are independent of other legislation, such as species-protection or animal-welfare legislation, therefore further regulations may apply. For example, if a dam is protecting a lodge when pregnant females and/or their dependent offspring are likely to be present, then any removal may be inappropriate on animal-welfare grounds. Similarly, the removal of dams which maintain water levels around an occupied lodge in the winter can result in the beavers abandoning the lodge and its attendant winter food cache. If water supplies cannot be managed by flow devices, then the most durable solution may be the installation of beaver-exclusion fencing. Where this solution is not feasible, then the capture and removal of beavers may be the only practicable alternative. However, beavers may soon recolonise if such areas can be continually accessed by dispersing beavers. In such cases longer-term management that permits a residential family is often more cost effective.

6.1.4.1 Dam-notching

This is the removal of a small amount of material from the top of beaver dam, usually by hand, to create a gap and thus increasing water flow over that section. It is most often associated with aiding fish passage but can also be used to lower water levels in beaver ponds behind the dam. It should be noted that, in active territories, beavers will often repair notched dams within 48 h. As a management technique, this can be quite labour-intensive, especially on a catchment scale. To obtain the maximum benefit from dam-notching, management efforts should be coordinated with fish migration (typically September–October: Parker 2013) or coupled with a flow device to manage the water levels in the long term. Dam size and longevity should also be considered before implementing dam-notching on a wider scale, as many dams are small, temporary in nature and/or will be breached or overcome during periods of heavy rainfall. It is most likely that only large dams should be considered for a seasonal dam-notching programme during years of low rainfall.

Certain management programmes, particularly in North America, seasonally implement the notching of active beaver dams just prior to or during fish spawning on traditionally occupied waterways. This is undertaken to reduce the risk of impeding the passage for adult fish (such as Atlantic salmon *Salmo salar*) attempting to access spawning areas during years of low water flow. Documented evidence of dam-notching as a regularly employed management technique in Europe is lacking, but recent Canadian examples exist. The Miramichi Salmon Association focuses on the conservation of Atlantic salmon stocks within the Miramichi catchment, New Brunswick. Similarly, the Gitanyow Fisheries Authority runs a Beaver Dam Breaching Programme in the Kitwanga catchment, British Columbia, to ensure the conservation of sockeye and coho salmon. To achieve this, beaver dams within the catchment were notched two or three times per week during the spawning migration period for these salmon species (Kingston 2004). In 2013, the Miramichi Salmon Association targeted >100 beaver dams for breaching in a watershed of 13,552 km² (Parker 2013). To aid in finding dams, and to target mitigation efforts and resources, aerial surveys were employed. In previous years, entire waterways were canoed. Dam-notching over a number of years has resulted in good populations of salmon and Brook Trout in catchments where they were previously low (Parker 2013). Dam-notching has been advised by fisheries as a useful practice for habitat-managers wishing to restore salmon populations (Canada: Taylor *et al.* 2010) but is labour intensive and may not be practical on a whole-catchment scale.

At present, dam-notching in Scotland may be undertaken on watercourses without prior authorisation from SEPA, using hand tools, rope or grapnels, providing such work is undertaken without causing pollution, including the escape of 'silty water' downstream (SEPA 2014).

6.1.4.2 Flow devices

The use of flow devices is a standard technique for lowering the water level behind beaver dams by influencing damming behaviour. Such devices were first developed in North America to provide a non-lethal response to the blockage of culverts by beavers. Several types of flow devices have been described, with varying terminology (e.g. Roblee 1987; Wood *et al.* 1994; Lisle 1996, 2001, 2003); and many less reliable copies are described on the Internet, so care should be taken, as the installation of inappropriate devices will lead to greater frustration and work in the long run. To maximise the value of flow devices, the design should be successful with respect to beavers but not overbuilt, while also being as long-lasting, durable and as maintenance free as possible.

Although initially this process will also address any associated issues with upstream flooding, beavers can readily create a series of subsidiary dams elsewhere in the landscape. If the objective is purely to reduce the water level at a specific point in a watercourse, then this technique has been shown to work well in parts of North America and Europe, provided that appropriately robust materials are used and installed in the correct manner. Periodic monitoring and maintenance may be required. The basic principle of this process is that the pipes going through the dam should be positioned so as to allow the water level to drop to the required depth, so that water is still present but at an acceptable level (as opposed to draining an area completely), with the pipe maintaining this level and taking away any overspill. If this design is incorrectly applied, beavers tend to respond by adding more material to the dam and/or building a new dam downstream of the flow device. This management method is generally considered to be effective in any depth of water if constructed correctly. In Bavaria, wildlife managers have found that dropping the water level behind the dam below a depth of ~80 cm has tended to encourage the beavers to construct another dam; however, experience in parts of North America has been that lower water levels behind dams can be successfully maintained with well-constructed flow devices. Such pipe systems have also been used effectively by experienced personnel to drain beaver ponds completely as an alternative to removing the beavers. Without the escape cover of pools, these beavers tended not to expose themselves in order to carry out dam-building. This method depends on beavers having other places nearby where they could dam without causing conflict. However, it is always a risk that dropping the water level too low may cause beavers to dam further downstream.

The installation of flow devices can be a very effective, relatively low-cost method for resolving beaver damming conflicts in comparison to regular road maintenance, dam or beaver removal (Boyles and Savitzky 2009; Simon et al. 2006). However, it should be noted that beavers may have multiple dams within a territory and may build tiered dam systems depending on habitat type. Failure rates for flow devices can be significant, and tend to occur in the first 2–12 months if they are installed by inexperienced personnel, are placed at inappropriate sites or are incorrectly designed (Czech and Lisle 2003; Callahan 2003; Lisle 2003). One study recorded a flow-device success rate of 87% at 156 beaver conflict sites in North America, with most failures resulting from insufficient pipe capacity, lack of maintenance and damming of the fencing associated with the flow device (Callahan 2005). The construction of new dams downstream by the beavers which negated the water level drop from the flow device was also considered a 'failure' in this study. Insufficient pipe capacity can be rectified by using a more appropriate size of pipe. High success rates have been demonstrated by other experienced practitioners. It should be noted that skill and knowledge will significantly influence flow-device success.

Figure 6.6 Piped dam. Note meshed inflow and metal stakes to keep pipe in place, and tree trunks adding extra weight (left), whereas in other situations the weight of the dam structure can hold a dam pipe, and a metal filter is not always required to work successfully. (D. Gow and G. Schwab)

Experience has shown that lack of success in flow-device installation is often a result of the pipes used being too small to handle large flows. As most dam creation occurs near the water surface in response to overspill stimuli, the pipe holds the water down during most of the damming season (low-flow periods). This results in a sufficiently low dam for the high flows to go over without causing significant flooding and potential damage to upstream infrastructure. Successful dam piping can be complex, requiring a number of environmental and design considerations, and consultations with experienced personnel are advised prior to construction to save resources and time and to reduce problems later on. It is important that the beavers are 'deceived' as to where the water is leaving their pool, as they are naturally attuned to search for leaks in or near to their dam structure. The sound and feel of water entering the inflow of the pipe must be masked by submergence, with its inflow located away from the dam (the concept of 'dam-leak separation'); however, the inflow of the pipe must be protected by a mesh filter, as beavers have reportedly plugged pipe inflows located >10 m away. If the beavers notice the inflow, they are likely to bury or otherwise to try to block the filter. The key is to prevent beavers locating the inflow, with good design, which allows the filter to be smaller and less expensive. Appendix C describes in more detail the types of pipe recommended for use in piped dams.

Advice sought from SEPA, through the Tayside Beaver Study Group (TBSG), regarding the placement of flow devices, has determined that these are likely to be classified as temporary structures and can be installed without prior authorisation under the General Binding Rules (GBR) under CAR. This would be on the condition that they are temporary structures (e.g. not fixed with concrete), and their installation was undertaken with minimal substrate disturbance (TBSG 2015). Additionally under CAR, SEPA considers that flow devices using boulders to secure any ancillary fencing can be installed without prior authorisation, as long as work is undertaken according to the conditions of CAR GBR14 (boulder placement in a river or burn). For any alternative approaches, prior consultation with SEPA is advised to determine any regulatory requirements (SEPA 2014).

Sewage plant – dam bypass flow device, Bavaria

After beavers repeatedly dammed an outflow ditch from a sewage settling plant serving the village of Saldenburg, Bavaria, management of their activity involved the use of standard flow-device piping that was inserted parallel to the dam itself. Machinery was used to construct a ditch into which the drainage pipe was placed to ensure that the inflow was positioned well back from the dam, with the outflow exiting lower down in the banking several metres away from the dam. This design has significantly reduced the flooding of the sewage plant while retaining suitable beaver habitat. No mesh basket has so far been used, as the beavers have not attempted to block the inflow, although this may be required in the future. The dam was not modified for piping, as there was a suitable breadth of adjacent bank, the sewage plant was easily accessed and machinery was available, resulting in this method being a more practical option.

Figure 6.7 Bypass dam flow device. Note inflow pipe on left; outflow pipe is marked by man in red jacket. Outflow of pipe exits banks several metres from dam. Photo below shows drained ditches after pipe installed – note dark lines of previous water levels. (R. Campbell-Palmer)

6.1.4.3   Modified field-drainage systems

Beaver dams in artificial field-drainage systems, particularly in very flat areas, are likely to be an ongoing management issue that may prove difficult to resolve even with, for example, repeated removal or flow devices. In Switzerland, field-drainage pipes have been modified so that, instead of draining directly into the main stream body, two collecting drains running parallel to the bank are dug. Any water is therefore collected from a series of field pipes, which are then directed to a point further downstream where they then empty into the main water body (e.g. Konolfingen, Switzerland). If such systems are coupled with riparian buffer strips as an adaptive management strategy, then beaver impacts should be significantly reduced. This system, however, has the potential problem that any raised ground water table as a result of increased water levels may impact on the drained land regardless. This system prevents the drainage lines from being blocked by any beaver-induced water-level changes, though any increase in the water level around the pipes may be an issue in itself.

6.1.4.4   Culvert management

Culverts are a common feature in developed landscapes. Although they vary in style and dimensions, they provide a basic drainage function for watercourses crossing beneath roads or through various infrastructure. The resultant backup of water can cause flooding and/or damage to the banks of the impounded structure. While the removal of material causing blockages can sometimes be effective, beavers can be persistent rebuilders, so that regular blockage removal may prove ineffective, laborious and not economically viable. The installation of oversize culverts where possible in the construction stage has also been proposed as a management solution (Jensen et al. 1999), although even wide culverts can act as a damming point in suitable habitat and if water is suitably shallow and slow moving. The structural design of culverts can also be important. North American research has demonstrated that box-type structures are less prone to blockage by beavers than pipes. This is thought to be because box culverts tend to constrict the channel width less than pipes (unless the pipe is buried deep into the channel, the width at the river bed is less than the maximum channel width), and the walls are smooth causing less turbulent and quieter water flow through them compared with corrugated pipes. Similarly, smooth walled culvert pipes are less prone to damming than corrugated metal culverts. Culvert guards which are designed at an angle to ensure either a flow-through of debris or to lower the possibility of blockage have proven effective against beaver blockage (Jensen et al. 1999), but these are likely still to require regular clearing.

Alternatively, a beaver-trapping programme (translocation or lethal) can be implemented, especially in areas with a history of repeated damming conflicts (Jensen et al. 1999; Czech and Lisle 2003), although this can be more expensive, especially over the long term (Boyles 2006). Additionally, draining an area completely may be a non-lethal way to encourage beavers to inhabit alternative sites, especially if the vulnerability of nearby properties is an issue. However, careful consideration must be given to the legality and expense of such actions, along with impacts on other wildlife and land-management practices. Therefore the most time-efficient, cost-effective or long-term management solution to reduce culvert blockage is often to install appropriate flow devices.

Unlike the installation of flow devices in dams, in culvert management it is important to discourage the attractiveness and usefulness to beavers of the area around the culvert. To do this, culverts must be protected from damming by a wood-frame structure (metal frames can be used in high-flow situations in which reinforcement is needed), which tends to be site specific in order to fit the topography in question. This is then combined with a similar flow device set as described above. Depending on location and surrounding topography the pipe can either be placed as low as possible so that water behind the culvert is kept shallow enough for the beavers to feel too exposed to use it (see Appendix C); alternatively the pipe can be set higher to allow the beavers to create a small pond near a road, if appropriate. This technique therefore gives control over where and how high the beaver dam can be. When a fence on, or in front of, a culvert is combined with an upstream flow device, it represents a strategy called a 'double filter system'.

A success rate of 97% has been recorded at 227 beaver conflict sites in North America when using this management technique, with most failures noted in this study resulting from a lack of maintenance, a lack of dam fencing or vandalism (Callahan 2005). It should be noted that, if beavers create a dam downstream of a flow device, this represents a response failure, and then this dam should be addressed as well. If insufficient pipe capacity is thought to be an issue, then an additional pipe should be added. Flow-device failure through lack of maintenance, or beavers finding holes, and/or vandalism can all be solved through repairing any damage and maintaining the structure more regularly. A well-designed and well-built modern flow device should eliminate many of these problems.

6.1.4.5 Fish passage

Fish passage through various flow devices has not reportedly been an issue, as long as filter mesh size is adequate. Larger openings can be cut if salmon passage is of particular concern, though holes larger than 16 cm may enable beavers to get inside the fencing and jeopardise any device. To ensure fish passage, pipes should be as low, level and straight as possible. Fish ladders and slotted box fishways (found on the Snohomish Pond Leveler) are designed as 'fish-friendly' management techniques which encourage fish

Figure 6.8 Fish ladders designed and built by S. Lisle. Pictured here on the left through a culvert, while on the right a fence has been purposely built so that a 'diversion dam' is created, i.e. beavers are allowed to dam along this fence line, in which a flow device has been constructed with a fish ladder (pictured on the right). (S. Lisle)

Figure 6.9 Slotted box fishway, attached to the end of a flow device. Key to this design is that the slots are ~50% the size of the pipe cross-sectional area so a pool with eddies is created in the wooden box and stronger current and areas of white water created at the exit to encourage usage by salmon. (M. Callahan)

movement by the creation of stronger currents and areas of white water at the opening slot to act as a fish attractant. The Clemson Leveller, which was initially designed to encourage fish passage through flow devices (Close 2003), is now generally accepted to be less effective than some of the more modern designs.

6.1.4.6 Dam-removal

Beaver dams are complex, interwoven structures. Removing dams may often stimulate beavers to rebuild again, usually felling additional trees in the process, which may exacerbate negative impacts. It may be more effective to allow a dam to remain, and to manage its size and the extent of resultant backwater. If the decision is made to remove a dam, then any discarded material should be placed above the high-water line to prevent it from being washed away and potentially causing blockage downstream. Such material tends not to be reutilised by beavers. The sudden and complete removal of dams, particularly using explosives, can result in the rapid emptying of the water from beaver ponds (e.g. WI DNR 2005), which can in turn result in the physical damage to or the death of fish (Niles *et al.* 2013). Additionally the sediment collected in the dam is washed downstream, causing further problems for fish and other aquatic species. Dams can be removed by hand (this can be more resource intensive and expensive but more controlled) or with an excavator where access is practicable, although the latter may result in damage to riparian habitat and more difficulty in controlling any sudden surge of silt and water downstream.

The removal of dams can also change the channel structure of a water body, often from a multiple-thread channel with pools and high complexity to a meandering

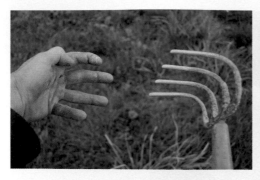

Figure 6.10 The best hand tools for beaver dam-removal. (S. Lisle)

Figure 6.11 Dam-removal by hand from a drainage ditch. (G. Schwab)

single-thread channel with no ponds (Green and Westbrook 2009). The effects of beaver dam-removal within a catchment have also been demonstrated to reduce the dominant riparian vegetation structure to dense scrub, significantly increasing the mean flow velocity and greatly increasing sediment yield (Green and Westbrook 2009).

Any removal of beaver dams using machinery may be currently undertaken without prior authorisation from SEPA, if this work is done according to the conditions of CAR GBR 9 (operating any vehicle, plant or equipment (machinery) when undertaking GBR activities 5, 6, 7, 8, 10, 12, 13 and 14) (SEPA 2014). If a dam is protecting a lodge, then particular consideration about its removal should be made, especially when heavily pregnant or lactating females and their dependent offspring are potentially present (April–September). Similarly, removing dams that are maintaining water levels around

Figure 6.12 Exposed beaver lodge following removal of dam further downstream (note previously submerged areas in foreground of photo). (R. Campbell-Palmer)

Figure 6.13 Removal of beaver dam with machinery. (G. Schwab)

an occupied lodge in the winter is not advised. Beavers are generally less active at this time, and smaller individuals may be less able to develop alternative living space at a time when water levels can fluctuate significantly. Beavers can also be dependent for their overwinter survival on sunken food caches which they assemble in the autumn. Any resultant lowering of water levels at this time can therefore result in the loss of their living space and food supply.

6.1.4.7 Discouraging dam-building

The absence or removal of woody material from the banks of a watercourse will not eliminate the possibility of dam-construction or the area being inhabited by beavers, and may cause additional issues such as bank instability. Where beavers attempt to construct

Figure 6.14 (Left) Electric fencing being used to deter beaver dam-rebuilding. (Right) Suspended gravel-filled plastic barriers, being used in combination with culvert protection. These may function to make the transportation of material more difficult and so discourage damming. (G. Schwab)

another dam on the same site, various dissuasive techniques have been attempted. The most effective of these short-term solutions is the use of electric fencing strung across the watercourse above the normal water level. Flashing lights (such as those used at road works) and ultrasound have proven to be temporarily effective in discouraging reconstruction at some sites (Schwab 2014). Other methods such as plastic barrels filled with gravel hanging from chains have been demonstrated as being effective on many occasions, whereas CD/DVDs suspended by cord above the dam have proven less so.

6.1.4.8 Grilles

Metal grilles are often used to protect small culverts. These are not management devices that control damming behaviour, but are included here for completeness of techniques employed. It should be noted that beavers may merely circumnavigate such management techniques and direct their activities elsewhere in a watercourse. Various designs have been developed to prevent beavers from blocking drainage pipes or culverts or from gaining access to an area. The main issue with these structures is that, while they are generally effective if robust enough, they can be blocked by beaver activity or by other detritus. As regular maintenance is essential, this technique is unlikely to be used widely outside of situations where routine checks are practicable (e.g. reserve staff). Grilles must be designed to ensure that they cannot be undermined by beaver burrowing and that gaps between the bars do not impact on fish passage. This can be accomplished by either setting the grilles into a concrete base or by constructing their main bars from rods (~1 m long) which are then driven down into the stream bed. Grilles should be fixed firmly into the banks or to the pipe itself with wire ties to ensure that beavers cannot burrow around the main barrier. There are many different designs for these structures, which vary according to requirements at individual sites.

Figure 6.15 Metal grilles to protect a culvert exiting a captive beaver enclosure. Note robust bank meshing and heavy gauge grille which is fixed to bank side and stream bank to withstand beaver digging around inflow. A sloping front tends to make clearance easier. Any grille must be cleared of debris by hand on a regular basis. (R. Campbell-Palmer)

6.1.5 Animal-welfare considerations

The impact of dam-removal on beaver welfare is generally only considered significant for those dams maintaining water levels directly associated with natal lodges or burrows: any sudden water-level drop during the breeding season may jeopardise the welfare and survival of kits. On animal-welfare grounds, removal and/or significant lowering of water levels should be avoided where a lodge/burrow entrance would be exposed during the breeding season when dependent young may be inside. Only about 10% of dams serve to conceal a lodge/burrow entrance (Hartman and Törnlöv 2006). The sensitive period for kit birth and emergence for any natal dam-removal is April–September. Removal of dams maintaining water levels around active lodges or burrows with an associated food cache from October to February should also receive greater consideration. Timescales could be reviewed upon consideration of local circumstances, particularly in harsh winters or regions experiencing shorter vegetation-growing seasons, e.g. northern Scotland in comparison to the Scottish Borders. Any resultant impact on winter survival should be avoided. Removal of such dams may encourage further tree-felling or relocation by beavers, or may present welfare challenges associated with food shortage. In harsh winters, welfare concerns over exposure, abandonment of food caches and increased energy requirements associated with relocation and shelter-creation would be of concern. Dams, especially more mature structures, are likely to provide refuges for other wildlife, so – if deemed necessary – should be dismantled with care. It should be noted that, if a dam stays in place long enough for it to be considered as 'mature' without causing conflict, then the need for its removal could be questioned. If heavy machinery is used, consideration should be given to its use in close proximity to other protected species and sensitive habitats. There are no welfare concerns directly associated with the removal of dams which do not create ponds covering the lodge/burrow entrance.

6.2 Burrowing and associated activities

6.2.1 Ecology

Beavers are strong, able burrowers and can readily excavate burrows, chambers and canals for shelter and/or access to food resources. Such structures may collapse and/or increase bankside erosion to varying extents depending on associated water flow and substrate type. Beaver burrows tend to be large and can end in sizeable chambers. Beavers will readily excavate burrow systems which begin underwater, and may be obscured by vegetation, so that their location may be difficult to determine. It is worth noting that any work causing damage or destruction of a lodge or burrow leading to deliberate or reckless disturbance whilst Eurasian beavers are occupying such structures is likely to be in contravention of the Habitats Regulations (SNH 2015).

6.2.2 Benefits

Burrowing and digging activities by beavers create microhabitats for other species in which to reside and forage. Such excavations increase the complexity and diversity of bank structures, which may also lead to increased meandering of watercourses. These features, in combination with the multilayered, dead-wood habitats which develop as a result of beavers' tree-felling activity, provide living opportunities for a host of wildlife.

A broad range of fish and aquatic invertebrates exist at high densities in these areas. These areas are readily utilised by a range of small mammals, while riparian predators such as otters or mink will exploit them as either temporary or breeding locations. Amphibians and aquatic reptiles such as grass snakes (*Natrix natrix*) are known to use the burrow systems of aquatic mammals as refugia and hibernacula. Adult burbot (*Lota lota*) have been identified as using submerged burrow cavities as lairs from which to ambush passing prey. Overall, such activities increase the diversity of a landscape by creating a mosaic of dynamic habitat types.

6.2.3 Issues

Human engineering over time has confined many watercourses between engineered banks or flood walls. Where embankments are set back from the main channel by a distance of >20 m, the impact of burrowing animals is insignificant. Where flood walls, dykes or other artificial structures exist <20 m from the water's edge, beavers can create burrows from the watercourse back into their main structure. Although the actual instances of beaver-generated burrows causing the collapse of engineered flood walls are few, the concern that the presence of beavers generates, has led some European water agencies to develop a range of remedial measures to counteract their activities (not only to mitigate for beaver activity, but also for muskrats, coypu, red foxes, badgers and rabbits). Beaver burrowing into flood banks has been reported in Tayside and has been linked to an instance of flood-wall failure and resultant flooding of crops (TBSG 2015).

In landscapes where networks of agricultural ditches are an essential land-drainage requirement, beavers will readily excavate burrows at the water's edge that can lead to bankside erosion and/or extend to the surface in adjacent fields. The secondary implications of equipment damage due to the collapse of beaver burrows, the delays caused to contractors' work schedules at harvest times, and the flooding of field edges and service roads may be much more significant and generally unacceptable to the majority of land-managers.

The cultivation of arable crops in the immediate vicinity of watercourses will often encourage beavers to burrow from the edge of the water into surrounding fields. In flatter landscapes, they can excavate canal systems out into the adjacent landscape. While their actual feeding impact on crops is relatively limited, the collapse of their burrows can result in operational delays and damage to agricultural vehicles or machinery. Burrow collapse may also pose an issue for livestock grazing.

6.2.4 Management options

6.2.4.1 Discouraging burrowing

Where old flood walls are already in place, the installation of effective mitigation is frequently physically difficult and expensive. When new flood walls are being constructed, the insertion of anti-burrowing mitigation can be incorporated as a standard feature. A number of effective anti-burrowing techniques (including those for beavers, badgers, rabbits, water voles, muskrats and coypu) are employed across Europe. These include the insertion of an interlocking system of sheet-metal piling through the centre of a main flood-bank structure or sunken welded wire fabric in smaller structures (this

Figure 6.16 (Left) Facing of flood banks with stones and mesh to deter digging (Bavaria). (Right) Sunken welded wire mesh (exposed by limited burrowing) to restrict any burrowing at a sewage pond in Bavaria. (G. Schwab)

is not used in flood banks). The facing of river banks with large rocks, concrete or stone gabions down to the bottom of the watercourse has also been effective. Beavers can easily gnaw through geotextiles (permeable membranes or fabrics designed for use in or on soil) and attempts to use these materials to deter burrowing have been ineffective. The Bavarian State Regional Water Authority has successfully implemented a range of flood-bank protection methods to mitigate against burrowing by coypu, muskrats and beavers (see Appendix C).

Although burrowing can generally be prevented by the installation of such 'hard' reinforcements, these options may be neither commercially viable nor ecologically desirable along extensive lengths of watercourse. Mitigation solutions such as the infilling of non-natal/residential burrows to prevent their use can be undertaken, though only after some form of monitoring for activity to ensure the burrow is unoccupied. After this, the original entrance holes can be covered with an additional barrier such as square

Figure 6.17 (Left) Flood-bank metal anti-burrowing plates being inserted in a flood bank for the River Danube; and anti-burrowing weldmesh. (Right) Specially targeted protection of a sewage filter-bed against beaver burrowing – this is cheaper than metal plates. (R. Campbell-Palmer)

welded wire mesh to prevent re-excavation. There are also examples of fisheries owners draining ponds to extend lengths of conventional livestock netting along the banks to prevent burrowing at the water level as a cheap but generally ineffective solution, especially if the shoreline is already pitted by beaver excavations or burrows. Plastic-coated mesh will corrode when it is permanently submerged, and lighter mesh gauges can be bitten through by beavers with relative ease. Chainlink has been used as it is often easier to work with, especially when trying to cover natural features, but this pliability means beavers can rapidly open large holes in this material. Mesh designs with unlocked joints can be easily distorted by pulling. At best, such techniques are dissuasive, and it is quite probable that other burrows will be excavated elsewhere. Where properties are located below fish ponds which could be breached by beaver-burrowing activities, the only practicable solution may be to remove any animals and then to exclude them from the area with appropriate fencing.

6.2.4.2   Artificial burrows

Artificial burrows have been created in otherwise netted riverbank slopes, specifically to encourage beaver use while protecting the remaining bank (Beck and Hohler 2000). It is important that such artificial burrows do not affect the stability of the river banks or any associated infrastructure (OFEV 2014).

6.2.4.3   Realignment of flood banks

The majority of all human–beaver conflicts occur within a relatively slim strip of habitat adjacent to freshwater habitats. In Bavaria, most beaver conflicts occur within the first few metres from water (see previous chapters). Although conflicts further away than this are possible, they are rare and are usually associated with an attractive food source. The creation of 10–20 m buffer strips of bankside vegetation, particularly wet woodland, will have the most significant impact on the reduction of potential beaver–human conflicts. While the purchase of the land behind an existing flood wall and the re-establishment of a new flood bank beyond the burrowing range of beavers is the most practicable long-term solution, it is an expensive option. However, if implemented correctly, this can improve flooding alleviation downstream while enhancing riparian habitat.

6.2.5   Animal-welfare considerations

Infilling or covering the entrances of burrows that are potentially still in use, especially during the breeding season in which dependent offspring may be present, would have serious animal-welfare implications. Extreme caution should be taken to ensure that no beavers (or otters) are present within the burrows. The potential for disturbance through the use of heavy machinery for beaver-mitigation techniques in and around natal burrows or lodges should be considered. If long stretches of water are faced with high gabions or smooth, steep concrete, there may be a welfare issue, as beavers (and other wildlife) may not be able to exit the water safely, if at all.

6.3 Foraging activity and associated management techniques

6.3.1 Ecology

Beavers can exist in both commercial hardwood and softwood plantations where deciduous tree species, shrubs or vascular plants are available. While Eurasian beavers do not commonly feed on evergreen tree species, they will occasionally ring-bark conifers or feed on their saplings in the late winter/early spring. Conifers can be felled for construction purposes, but this tends to occur only if few broadleaf trees are available. There have been no reports across Europe of nationally significant economic or ecological damage to woodlands caused by beavers (Reynolds 2000).

Beavers display regular routines and feeding patterns, with well-worn foraging trails and canals being easily visible. This characteristic can be used to deter them in some situations where their presence is unwanted. If possible, felled trees should be left *in situ*, as beavers will continue to feed on their side branches and bark. They will also utilise these materials for construction. The removal of such trees is likely to encourage further felling. If a felled tree must be removed as it is causing an obstruction, further felling can be discouraged by the methods outlined below.

6.3.2 Benefits

The selective foraging activities of beavers have great ecological benefits, and can create diverse habitat types and structures. In riparian forests, the felling of trees creates open

Figure 6.18 Long-term beaver-grazed lochan riparian margin in Norway (~30 years of beaver occupation). The growing season locally is <100 days/year, so that the beavers eat a predominantly woody plant species diet and the vegetation grows more slowly than would be expected in Britain. Note trees still present and plant species diversity in the foreground, where the beavers have created and maintain the bushy 'beaver lawn' in the centre of the shot. (D. Halley)

patches in the canopy, thereby changing the forest structure, composition and even vegetation type as varying plant successional stages occur. Biodiversity can be further enhanced through repeated and selective grazing. Tree-felling creates fallen dead wood, which is vital for forest health and to provide habitats for saproxylic species especially beetles, and hence to support a host of associated species.

6.3.3 Issues

The consequences of tree-felling by beavers into or across a water body will vary greatly in accordance with its size, depth, width and associated water flow. Woody debris in river and stream systems is a naturally occurring feature that large commercial users of water, such as hydroelectric projects, cater for as a matter of normal process. If the watercourse is very large, such activities will have little impact. Most large trees will remain *in situ* where they fall and will not present the risk of hazard elsewhere, unless a significant flood event occurs. Extracting trees from a watercourse once they have been felled or undermined by beavers can be a specialised and expensive undertaking. The felling or ring-barking of large specimen trees in certain landscapes could generate a sense of cultural loss. In large-scale forest blocks, beaver foraging is largely an insignificant issue, with impacts generally more significant through flooding of tree stands. In smaller, community forests where particular trees are desired by individuals for carpentry or firewood purposes, their loss can have a greater impact on local stakeholders. See also section 5.5.1 on agriculture.

6.3.4 Management options

6.3.4.1 Individual tree protection

Individual wire-mesh tree guards are an effective method of protecting trees from beaver browsing. Although standard 1 × 2 inch (2.5 × 5 cm) weldmesh tree guards are effective, versions constructed from light mesh types such as 1 inch (2.5 cm) 'chicken'

Figure 6.19 Trees can be protected easily with mesh or anti-game paint. (D. Gow and R. Campbell-Palmer)

Figure 6.20 Apple trees wired with chicken mesh after beaver gnawing had occurred. These trees will probably die as management techniques were carried out after significant gnawing had occurred. In such circumstances, especially for valuable specimens, tree grafting could be implemented. (R. Campbell-Palmer)

Figure 6.21 Ineffectual tree protection in which mesh diameter is too large to prevent beaver gnawing. These dimensions of weld mesh can be utilised in tree protection if the diameter and height of the mesh cylinder is increased. (S. Lisle)

or ½ inch (1.25 cm) 'rabbit' wire are also effective in environments where beavers have an ample alternative food resource; however, if not, it should be noted that this type of wiring is easily pulled down by determined beavers. Beavers will commonly ignore trees with guards of this type, providing they cannot push under or pull down the mesh to access the trunk or side branches. Any meshing should not impact on tree growth. Where amenity trees are to be protected but mesh guards are considered unsightly, then anti-game paint, e.g. Wöbra®, has proven effective against beavers. This product was initially designed to protect against Red Deer browsing, is translucent when dry, and contains grit which many browsing mammals dislike chewing on. A cheaper, 'home-made' version consisting of non-toxic exterior paint and masonry sand will also deter beaver feeding. Both these abrasive paints have been successfully trialled in Scotland (TBSG 2015). In areas where significant snow lie is likely, note that beavers can and do stand on top of the snow and fell trees; the height of any protection should be adjusted accordingly.

While felled trees have the potential to cause incidental damage where they fall on fences, power lines, buildings, transport routes and even vehicles, the frequency of these events has in practice turned out to be rare – though, when they do occur, they can be significant in terms of disruption, cost and are often over-reported by the media.

6.3.4.2 Deterrence techniques

Practical deterrents that impact on new activity by beavers, before they have become established, can be useful under certain circumstances, e.g. to deter the rebuilding of dams in a specific location or to discourage beavers from using a feature such as a garden pond, especially when their presence is still at an exploratory level. Again, techniques such as flashing road lamps and ultrasound have been reported as being successful on a temporary basis. The presence of free-running dogs in private gardens has also been reported as an effective deterrent.

A number of commercially available chemical deterrents are available on the market, particularly in North America. These make varying and potentially outlandish claims over their effectiveness. The application of predator odours has been determined to reduce beaver foraging, particularly otter scent (Rosell and Czech, 2000). Overall, although some substances may prove effective in some circumstances, in the long-term they are generally considered ineffective as they have to be reapplied repeatedly which can be expensive and labour intensive (Wagner *et al.* 2000).

6.3.4.3 Electric fencing

Electric fencing can be employed to deter beavers, provided it is positioned correctly. It is portable, easy to install and can be relocated rapidly in response to any change in activity. Electric fencing is widely used by many landowners, for example to prevent beavers from feeding on arable crops. It can also be utilised as a temporary measure to deter beaver activity in a variety of situations, e.g. dam-height restriction or deterring beaver movement between a water body and an ornamental pond or garden. If beavers are able to burrow underneath the fence and re-emerge in fields, this technique may be less effective, but it is often an effective deterrent against crop feeding.

6.3.4.4 Exclusion fencing

Exclusion fencing can be used to protect an area more permanently, e.g. a stand of trees, or to discourage colonisation of small streams or the use of ditches. If colonisation of a new ditch or small stream is undesirable, it may be possible to prevent or delay immigration by

Figure 6.22 Double-strand (~15 and 25 cm from ground level) electric fencing used to deter beavers from feeding on sugar beet (Bavaria). Use of electric fencing is seasonal and limited to edge of crops within range of nearby watercourse. Note use of herbicide to prevent plant growth from shorting electric fencing – however, given the wider impacts of pesticides on the environment this is not recommended. (R. Campbell-Palmer)

turning beavers back near the confluence point(s). This process involves the construction of a beaver-deterrent fence (using what may also be called a Swept-Wing Fence™ that crosses the ditch/stream/burn with 'wings' back towards the stream), which aims to turn any beavers exploring the area around to double back upon themselves as opposed to navigating around the fence. The longer the wings, and the more they confuse the beaver by sending them in the 'right' direction, the more robust (and expensive) the system tends to be (L. Skip, personal observation). In more mountainous or rocky terrain, it may be possible to select a point along the watercourse where the fence can tie into a natural barrier. A flow device (pipe system) at some point on the upstream side of the fence(s) may additionally be required, as is it possible for beavers to get inside an 'exclusion zone' and incorporate any fencing across a water body into a potential dam.

This management method exploits the general reluctance of beavers to move far from water. In France, the authorities have built special 'beaver ladders' which make it possible for beavers to bypass human-made dams and other barriers in France's heavily regulated rivers. In Norway, beavers have been recorded travelling overland to access water bodies above large hydroelectric plants. The ability of some beavers to travel longer distances than can be generally expected should not be underestimated; this motivation varies

Figure 6.23 (Left) Beaver-deterrent fencing, more robust in the stream to secure against beaver digging and flood debris; and larger mesh size to allow fish and otter passage. (Right) More robust fencing used at a beaver demonstration project. (Scottish Beaver Trial and R. Campbell-Palmer)

between individuals and may be dependent on available resources and reproductive status, also the motivation will increase with increasing populations.

There are well-established designs for permanent beaver fencing. These are most effective when they are positioned well back (~20–30 m) from the edge of a watercourse. Fencing should extend underground, or have a collar at ground level with a 90° angle and an extension of around 40 cm towards the beavers' likely direction of approach. This collar should be pegged into the ground to allow vegetation regrowth through this mesh to form a secure mat. These mesh collars discourage beavers from digging along a fence line. Any trees within falling distance on the outer side of the fence line should be removed, coppiced or protected from gnawing to ensure that beavers do not fell them onto the fences, thus damaging and potentially breaching such protective fencing. If the fence is constructed from high-tensile wire, any minor trees or limbs are generally held on its upper strand without significant collapse. Although expensive to construct, these fences may have a role in protecting commercial orchards, biofuel plantations or arboreta.

The most effective known design for fences of this type is based on the systems used for retaining captive beavers. This design can be accompanied by the use of an electric 'hotwire' fence (with care) and/or sunken fencing. Details of such fencing have been described in Campbell-Palmer and Rosell (2013) and in the Natural England advisory notes (draft) included in Appendix D.

6.3.4.5 Land-purchase schemes

Government compensation schemes are common in some European countries to incentivise the toleration or acceptance of reintroduced species on private land. These programmes can be quite successful for the re-establishment of reintroduced species or the protection of isolated populations of protected species (Cope *et al.* 2003; Naughton-Treves *et al.* 2003). While programmes of this type are generally popular with wider society, the opponents of such schemes question their long-term cost implications together with the ethical basis of providing financial incentives to landowners for adopting a more wildlife-friendly approach (Bulte and Rondeau 2005). In Bavaria, the purchase of land to enable a resident beaver population to establish or expand a wetland environment is a developed part of beaver management, for example, using local authority or central state funds. The application of planning compensation agreements whereby developers fund the creation of wildlife habitats to compensate for their activities elsewhere has also resulted in the creation of beaver-generated wetlands on the edge of urban areas such as the village of Auerbach in Bavaria.

6.3.5   Animal-welfare considerations

A few deaths have been reported in captive collections, involving beavers biting 'hotwire' fences where their front teeth became locked behind the wiring. The use of such fencing should be carefully considered and incorporate monitoring via remote camera traps should be incorporated. Any deterrent should be situated so that a beaver is able to remove itself.

6.4   Animal management

The techniques described in the preceding section focused on the management or mitigation of beaver impacts. This section considers the management of the animals themselves. The perceived need for, and methods of, regulating beaver populations vary greatly across Europe, from hunting quotas in Norway (frequently unlimited due to demand for hunting being below the rate of natural increase in many river systems) (Parker and Rosell 2003), to trap-and-remove by employed staff or trained volunteers in Germany (Schwab and Schmidbauer 2003). In countries where beaver populations are still recovering, they are usually fully protected, and mitigation through non-lethal management methods prevails. The potential population management strategies for Britain have yet to be fully investigated and will be subject to further discussion and legal considerations should beavers be formally reintroduced. It should be noted that deliberate capture, disturbance, killing (i.e. dispatch after live trapping) and the possession (including carcass or parts of), control and transport of Eurasian beavers is likely to be contravention of the Habitats Directive and require a licence.

Where beaver populations are reintroduced or re-established, their presence in developed landscapes will ultimately require some system of management. In western Britain, where the production of livestock predominates, limited human infrastructure exists, and the landscape is dominated by wooded river systems and upper valleys or heavily grazed highlands, the presence of beavers is likely to result in relatively limited conflict. In eastern or lowland landscapes, where arable production increases in importance, and greater densities of people are present, then over time the presence of a substantial population of beavers will result in the same issues which have arisen in other developed landscapes such as Bavaria.

There is no effective methodology for excluding beavers from large areas of suitable habitat, although freshwater bodies with rocky or concrete banks and no attendant food resource are relatively hostile environments. The creation of such environments specifically to exclude beavers will not always be a practical or desirable solution. In cases where beaver conflicts cannot be suitably managed, as costs are too high or the potential impacts too great, the removal of their presence through a process of trapping and translocation or culling may be the only practical solution (Rosell and Kvinlaug 1998; Parker *et al.* 2006a). Although at the time of writing the beaver is not listed as a European Protected Species in domestic legislation, this is likely to change if beavers are formally reintroduced to Britain and there will be a clear requirement for a licence to trap or kill beavers.

6.4.1   Ecology

Any trapping effort to remove beavers from an area should seek to ensure that no dependent offspring remain. Juvenile beavers, for at least their first year of life, are reliant on their parents and older siblings for shelter, protection and food provision. They also rely on the communal body warmth of larger individuals in winter. The trapping and relocation of heavily pregnant or lactating females should be avoided. Any trapping programme must recognise that a repetitive process of trapping and monitoring (pre- and post-) will be required to remove beavers from an area completely. Beavers commonly display varying levels of bait and trap shyness, with subadults and males tending to be more easily trapped (Schulte and Müller-Schwarze 1999; Müller-Schwarze 2003).

As beavers are a highly territorial species, fighting can inflict serious wounds and even death, so care must be taken to ensure that any translocated individuals are not released into the territories of non-related animals. Beavers from different family units should never be mixed in the same transport crate. Ideally, pairs or family groups should be released together in available habitat. Any translocation should follow best-practice guidelines, including IUCN (2013) and Best Practice Guidelines for Conservation Translocations in Scotland (National Species Reintroduction Forum 2014).

6.4.2   Benefits

The live capture and translocation of beavers has been widely used to re-establish the species in continental Europe. Initially, when scarce, beavers were obtained from whatever donor populations were available. Later, a more targeted process involving the capture and movement of 'problem' beavers was developed in Bavaria as a successful technique combining mitigation with reintroduction. Since the early 2000s, approximately 1000 individuals have been captured and moved under licence to countries such as Croatia, Romania, Hungary, Serbia, Romania, Croatia, Bosina-Herzegovina and Mongolia. These releases, which typically focused on the capture and movement of many family groups for simultaneous releases on a single river system, have proven highly effective. Stable and expanding beaver populations have been established as a result of this project in all of the above states (apart from Mongolia). This is, however, a specialised process that requires the cooperation of a broad range of partner organisations and governments. Properly experienced field staff and vets, and facilities for holding, capture and transport, are also essential.

In continental Europe the range expansion of established beaver populations is narrowing the remaining options for projects of this type, and currently in Bavaria ~1300 beavers are now culled annually. In Britain, translocations on a much smaller scale could prove to be an effective option for dealing with 'problem' animals. Beavers are poor colonisers of water bodies overland, and this option, under licence, could for some time reduce (though not eliminate) the requirement for culling or harvesting through hunting, in environments where the presence of beavers is undesirable.

6.4.3   Issues

Currently there is little beaver-trapping equipment and expertise in Britain, so any future translocation or culling programme would probably need to consider a greater investment in appropriate trapping management. The trapping and translocation of 'problem' individuals is a useful but ultimately limited management tool. It is generally resource-intensive, as a number of criteria would need to be satisfied for its implementation, including legislative compliance, suitable equipment, assessment and identification of release sites, monitoring at trapping and release sites, and appropriate health-screening of individuals as required.

Based on the evidence from reintroductions elsewhere in Europe, and of free-living beavers in Britain, any populations which are not eliminated swiftly will usually establish and expand. Eurasian beavers are now re-established in much of Europe (bar their western and southern former ranges), and there is no real prospect of translocating beavers from Britain as a means of ameliorating local beaver conflicts. If internal translocation is not politically possible or when its potential is fully exhausted, then management through culling when irresolvable conflicts arise will be the only practicable option over time.

6.4.4   Management options

6.4.4.1   Wildlife rehabilitation

If the Eurasian beaver is reintroduced to Britain, it is possible that injured beavers will from time to time be encountered by members of the general public and by wildlife rescue organisations. Orphaned juveniles, or individuals perceived by the public to have been abandoned by their parents, are likely to be handed in to local wildlife rescue centres – though it is very difficult to conclude if any discovered kit has truly been abandoned by its parents, and it is more likely they are sick or injured if they can be handled. Hand-rearing of beaver kits is time-consuming, and they require a suitable diet. Any organisations receiving beaver kits should seek specialist advice. Dispersers or adult

Figure 6.24 Young kit handed in to a wildlife rehabilitation centre (Bavaria). (G. Schwab)

beavers can be seriously wounded in territorial fights. This can create deep penetrative wounds which often seal at either end and can become septic. Such individuals may lie up in dense vegetation or under bridges, for example, where they can be discovered by dogs or people. Adult beavers can defend themselves effectively against pet dogs. In Bavaria and Norway, where beavers live in areas where recreational dog-walking is common, such encounters are rare but not unknown.

Few wildlife rescue centres currently have appropriate facilities for beaver care, though modifications could be made to otter or seal facilities. The Scottish SPCA is aware it is likely to receive injured beavers in future and is considering how it would accommodate future requirements (C. Seddon, personal communication 2014). Beaver captive care can be relatively straightforward, though their veterinary and hand-rearing requirements are less well documented, and training and information dissemination among wildlife rescue or veterinary facilities may be appropriate. A greater potential issue is the future of rehabilitated beavers that reach the stage of possible release. Although under the Habitat Regulations (as amended) a licence would not be required for possession and transportation for rehabilitation purposes, it would currently be considered illegal under the Wildlife and Countryside Act 1981 (as amended) to re-release any rehabilitated beaver into the wild without a licence. Further consideration must be given to their future prospects by wildlife-rehabilitators who encounter these individuals, and/or a legal framework established to enable rehabilitated beavers to be returned to the wild. An additional consideration would be the welfare of any released individuals, particularly in relation to the identification of suitable release points. Many wildlife hospital organisations have an ethical opposition to the euthanasia of healthy animals. Without any option for the re-release of rehabilitated beavers, alternative solutions for their long-term care would be required. It is unlikely that many zoological and wildlife collection facilities would provide an outlet for such individuals.

6.4.4.2   Fertility control

Sterilisation of beavers may be an option to reduce breeding in an area with stable families instead of culling, as breeding pairs are likely to remain in their territory. Sterilisation has been trialled as a management technique but its use is not widespread (Brooks *et al.* 1980; Pizzi *et al.* in preparation) as significant resources are required by way of trapping animals and performing veterinary procedures under anaesthetic.

In the Rhineland-Palatinate region of Germany, free-living North American beavers are being sterilised as a result of public objections to the concept of lethal removal. This project, which has been funded by central government, has resulted in ~60 beavers being sterilised over a 6 year period. This management strategy was determined to be a positive option on the basis that, while it controlled the spread of the non-native North American beaver, it allowed the existing population to produce positive ecological benefits and supported the return, over time, of Eurasian beavers (S. Venske, personal communication 2015). While appropriate in this special case, sterilisation is too expensive and labour-intensive to be practicable on a large scale (Schulte and Müller-Schwarze 1999).

Beavers of both sexes can be permanently sterilised surgically. While standard veterinary sterilisations (ovariectomy or ovariohysterectomy in females, and castration in males) have been performed on beavers, these processes can result in adverse effects such as behavioural alterations and family-structure collapse when the gonads are removed. It is preferable, if performing surgical permanent sterilisation, to perform tubal ligation

 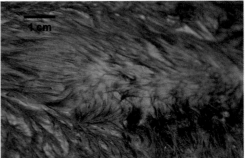

Figure 6.25 (Left) Fur clipping and sterilisation scar (Bavaria). (Right) Laparoscopic sterilisation scar (no fur clipping). (Zoological Medicine)

in females and vasectomy in males. This preserves the normal gonads and production of sex hormones, with retention of normal behaviour and social and hierarchical integrity. Both tubal ligation and vasectomy can be performed in a minimally invasive manner by laparoscopy (also referred to as keyhole surgery) via two 3–5 mm incisions (Pizzi *et al.* 2012, 2014). This approach can be performed without the need to clip any fur, with the advantage of minimal effect on waterproofing and thermal regulation.

This method has been employed in the sterilisation of North American beavers in parts of Germany, with recovery times of a few days. Successful recaptures of sterilised individuals in different locations have been recorded, with no negative impacts reported to date (S. Venske, personal communication 2015). Beavers have been returned to their free-ranging aquatic environment within 12 h of surgery, with no adverse effects after long-term monitoring. If either the male or the female from an adult breeding pair is sterilised, then other family members do not tend to breed for as long as the sterilised beaver remains in the family unit (Brooks *et al.* 1980). Female beavers display declining reproductive output as they age, with individuals able to survive and defend their territory for several years past their last successful reproductive attempt (Campbell *et al.* in press). Declining fertility is therefore expected in individual family groups where the dominant female remains unchanged.

Figure 6.26 A North American beaver being re-released after sterilisation (Rhineland). (S. Venske)

6.4.4.3 Live trapping

If beavers are reintroduced or permitted to spread from existing areas, then trapping and relocation are likely to be required, particularly in circumstances in which protective measures are too expensive or impractical to implement (Schwab and Schmidbauer 2001). Beaver live-trapping methods vary between countries, and there are several designs of beaver traps and types of trapping techniques (Rosell and Kvinlaug 1998). In North America and parts of Europe, live snares and 'suitcase-type traps' such as Hancock and Bailey traps are regularly used (e.g. McKinstry and Anderson 2002), although there are welfare concerns associated with these trap types – they afford no cover from the elements to any captured beavers, and, if dislodged, can present a potential drowning risk. Live trapping from a boat is used in Norway (Rosell and Hovde 2001). The use of spotlights and specially designed 'landing' nets makes it possible, if individual animals can be identified, to target specific individuals and avoid multiple recaptures of untargeted animals, thereby reducing both stress on the animals and trapping time and effort (Rosell and Hovde 2001). This is a technique requiring experience and training, and the nature of the water bodies or the conditions may not permit boat-trapping. In particular, the water body needs to contain areas of water shallow enough for the trapper to jump into (not deeper than waist height) but deep enough to navigate a boat. The base bed of the water body also influences success with this technique, with excessively rocky substrate hindering capture. Cylindrical 'Belarus' beaver traps are used in Russia and Mongolia (Müller-Schwarze 2003). Capture rates were previously recorded as 11% for Hancock traps, 16% for Bailey traps (Hodgdon 1978) and 21% for box traps (Koenen *et al.* 2005). Live boat-and-net trapping by experienced personnel can achieve a trapping rate of approximately one beaver per hour on river systems (F. Rosell, personal communication 2015). Trapping mortality has been reported as 4% in live snares, 1% for Hancock traps (McKinstry and Anderson 2002), 0% in box traps (Koenen *et al.* 2005), 1–2% in Bavarian traps, and 0% for the boat-trapping method (F. Rosell, personal communication 2015). At the time of writing, there are no approved spring traps for beavers in Britain, so cage trapping is presently the only option.

Bavarian beaver traps, which have been developed over a number of years, are probably the most effective and accessible trapping method in Britain. They are designed to afford any captured beaver a considerable degree of protection from the weather, while reducing trap-inflicted injuries. When constructed from aluminium, these traps are comparatively light and easy to manoeuvre in the field. Care must be taken with trap placement and position of bait, and to ensure regular servicing. Any beaver-trapping project must be aware of likely water-level fluctuations in the area which may endanger any captured beavers, and must also ensure that traps are set away from likely locations of human interference. Any set traps should be checked at least once a day when in operation, ideally being set in the early evening and then checked the following morning. Although these traps also have the ability to capture other species, incidental captures are low. In five years of use at the Scottish Beaver Trial, one badger and two pine martens were incidentally trapped unharmed; and one Pheasant was incidentally trapped during the monitoring project for the Tayside beaver population (Campbell-Palmer *et al.* 2015). Any trapping should cease when heavily pregnant females and/or dependent juveniles are potentially present (April–September).

In order to increase capture success and reduce capture effort, a good knowledge of beaver behaviour and of their habitats is vital. Experience of using trail cameras to identify the number of beavers present in a colony and their response to pre-baited traps

Figure 6.27 Bavarian beaver trap with camouflage painting, ready for dispatch (~L 175 cm, H 60 cm, W 60 cm). (G. Schwab)

Figure 6.28 Lamp spotting for beavers (Norway), with the landing net lying horizontally in the middle of the boat. (Scottish Beaver Trial)

is useful. Appropriate capture methods may vary according to habitat type, season and the experience of the personnel involved. Often a combination of capture techniques may be required (see Rosell and Kvinlaug 1998; Rosell and Hovde 2001), especially if habitat and water conditions do not permit boat or hand-capture methods. Boat capture is an inappropriate technique in small water bodies, sites with poor water clarity, and areas with shorelines with many submerged obstacles and/or unstable substrate. Specifically designed beaver traps (such as Bavarian beaver traps) can then be used, and should be set along forage trails, near dams or canals, or near feeding stations. Beavers may also be tempted into traps by repeated baiting with appropriate food or castoreum. Castoreum has also been used as a beaver attractant for centuries. Sweet plant foods such as apples, carrots and sweet potatoes have successfully attracted beavers to traps (Scottish Beaver Trial; Bavarian Wildlife Authorities), as have cut branches of Aspen or Birch (Koenen *et al.* 2005).

Any programme of beaver-trapping in Britain must ensure that the relevant Statutory Nature Conservation Organisations (SNCOs) have been consulted and that any appropriate licences have been obtained.

6.4.4.4 *Translocation*

Under appropriate licensing, translocation could provide a relatively cost-effective source of beavers for recolonisation projects, if the health and genetic status are considered favourable. While the trapping and translocation of 'problem' beavers is a viable management tool, at least in the short to medium term, as suitable habitat becomes occupied it will provide a more limited solution. Translocation requires a combined process of legislative compliance, identification of suitable release sites, health-screening and post-release monitoring. In Scotland, the Scottish Code for Conservation Translocations has been published, and any conservation translocation is now expected to follow this agreed approach (NSRF 2014). As beavers are highly territorial, care must be taken to ensure that they are not released into territories already used by other beavers. Ideally, pairs or family groups should be released together. Potential relocation sites should be identified in advance, with landowner permission secured. It would be prudent to ensure that a common standard is developed for this process, to ensure best practice and high standards of animal welfare.

The transport of beavers over long distances should only be undertaken in specially constructed transport crates. There are various designs for these, but the best are built from aluminium or wood overlaid with heavy-gauge weldmesh or sheet metal. Collapsible carrying handles on the tops of the crates allow two people to carry them and make it easy to stack other crates or equipment on top. Small sliding inspection hatches at the far end to enable feeding and checking of individuals during transport is recommended. Adult and subadult beavers should be transported in separate crates, while yearlings and kits can be crated either together or with a parent. Groupings of this type should not be confined together for long periods and must have enough space for free movement. Transportation must be in well-ventilated vehicles or on trailers to ensure that the beavers do not overheat. A deep layer of straw/sawdust must be provided for each crate, and sweet apples should be provided for moisture. Any transported animals should be kept in darkened environments (with appropriate ventilation) without excessive noise or sudden light and movements. Where individuals from different families are being moved in the same vehicle, care must be taken to ensure that their paws or teeth cannot come into contact with other beavers, as serious injuries from bites can easily occur. Beavers from different families should never be mixed in the same crate, as they are likely to fight savagely.

At higher beaver population densities, this management option may become less practical. The Department for Environment, Farming and Rural Affairs (DEFRA) has a voluntary code, www.defra.gov.uk/animal-trade/files/Code-of-Practice_Beavers1.pdf, for those importing beavers into England and Wales. This particularly recommends that wild-caught beavers are sourced from regions free from *Echinococcus multilocularis* or

Figure 6.29 (Left) Modified wheelie bin used to transport beavers short distances to a release site. (Right) Internal view of specifically designed beaver transport crates (~1 m × 60 cm × 55 cm), conforming to IATA regulations. The deep tray should be filled with absorbent substrate such as sawdust and hay, as beavers tend to urinate in crates and traps, which should be taken into consideration if castoreum collection is an issue. (G. Schwab and R. Campbell-Palmer)

Figure 6.30 (Left) Beaver released following health and genetic screening, River Earn, Tayside. (Right) Beaver release at the Scottish Beaver Trial. (H. Dickinson and G. Goodman)

are captive-bred animals that have not been exposed to foxes and domestic dogs, the definitive hosts of this parasite.

6.4.4.5 Culling

Nordic research has addressed numerous questions associated with beaver-hunting, including differences in the susceptibility of sexes and age groups to being shot, with female adults more likely to be shot during dusk in spring (Parker *et al.* 2002b), the effect on both breeding time (Parker *et al.* 2006b) and population size (Ermala 1997, 2001; Hartman 1999; Parker *et al.* 2002b; Parker and Rosell 2012, 2014), when the hunting season should be closed for animal-welfare reasons (Parker and Rosell 2001; Parker *et al.* 2001b; Parker *et al.* 2006a), the killing efficiency and effect on pelt and meat quality of different rifle ammunition (Parker *et al.* 2006b), and the overall role of hunting in the management of beaver populations (Hartman 1999; Parker and Rosell 2003, 2012). In the USA, hunting rights belong to the public, and therefore state-employed game wardens are usually responsible for undertaking beaver-damage control. If the Eurasian beaver is formally reintroduced to Britain, it will most likely be considered as a European Protected Species, and therefore any future hunting (hunting here referring to the use of guns, as opposed to hunting with dogs) could only occur under the appropriate derogations unless the Habitats Directive is revised and it is determined they are no longer a rare species. If translocation is not possible, then management of beaver populations through culling in catchments when irresolvable conflicts arise will be the only practicable option. Hunting is widely employed in Northern and Eastern Europe, both as a recreational activity and for managing populations. This process offers a number of attractions: it is flexible and if appropriately structured, it can be almost bureaucracy- and cost-free (it can generate income in some countries); and it can contribute to reducing conflicts over land-use between humans and beavers. However, Britain lacks a tradition of beaver-hunting and the concept itself may be controversial. It is highly unlikely that beavers would ever be managed through harvesting for fur, given the public unacceptability of fur in the UK. Additionally, there is no significant market value in beaver pelts, and for this reason it would be uneconomic (Balodis 1994).

Lethal beaver control is best expressed to wider society, conservation and animal-welfare groups as a targeted response to the activities of problem individual beavers or family groups. Most hunting is undertaken with rifles, although shotguns are permitted in

Figure 6.31 Beaver-skinning, Norway – predominantly for meat, but pelts are still used in some areas. (E. Solberg)

Figure 6.32 Beaver with forest mushrooms, Norway. (D. Halley)

some jurisdictions. Hunting for beaver works best in those countries where the landowner controls the hunting resource on his/her land. In places where the landowner does not control the harvest, public opinion can limit the practical application of this approach given the unpalatability of hunting to the general public and strong reactions against wildlife culling, e.g. hedgehog culling in the Western Isles of Scotland, and badger culling for the control of bovine TB. At present, beavers are protected in Europe by the Habitats Directive, but a number of countries have derogations (allowing beaver-hunting/culling). Management could be decided wholly at a national/regional level. In countries in which hunting is currently permitted, beavers can be removed by landowners at their own discretion and without subsidy. In modern conditions, this seldom constitutes a threat to beavers at a population level. Experience shows that many landowners are more willing to tolerate beavers under such conditions than when they are perceived as 'wards of state agencies'. However, for hunting to be a consideration in Britain, it is most likely that it

Beaver harvest management in Norway (Parker and Rosell 2012)

From total protection status in 1845, beaver populations in Norway have seen a number of management schemes implemented to maintain populations in order to enhance biodiversity, produce a harvestable surplus and reduce human–beaver conflicts. The Norwegian 'Wildlife Act' details the hunting and trapping seasons and the methods permitted. This hunting largely takes the form of spring shooting, although some winter trapping also occurs.

Central to beaver management in Norway is harvesting, both to ensure conservation and to manipulate and exploit beaver populations. Quota regulations have been successfully implemented since 1955 and involve wildlife managers determining whether local populations can be harvested, through information gathered from landowners, hunters and counts of occupied lodges. An annual harvest quota is established on a township scale, then divided among landowners according to the total areas of beaver-occupied habitat they own. Typically, landowners may group together to gain increased quotas, which they may hunt themselves or sell to hunters, thus generating an income.

would be on a small scale, at limited sites on a recreational basis in which hunters, with the permission of landowner, remove beavers for the experience, potentially paying for the opportunity. It is unlikely to be an effective population-management strategy across Britain.

6.4.4.6 Humane dispatch

Successful beaver management will eventually require humane lethal control initially through the identification of problem animals, or the removal of beavers to prevent colonisation of predetermined 'beaver-free zones', or to achieve annual culling quotas (realistically not a perceived management requirement for a number of decades), before beaver populations are widely established. If beavers are to be dispatched by shooting, then certain factors should be considered to ensure that dispatch is humane. Currently, beavers can be shot without a licence in Britain, provided landowner permission is granted, and firearm and animal-welfare laws are complied with. However, this situation is very likely to change if the species is formally reintroduced.

Where hunting is not considered appropriate, the key to successful beaver-management on a landscape scale lies in the specific identification, capture for translocation or humane culling of the family or individuals which are causing a problem. In such instances when lethal culling is deemed necessary and live trapping followed by euthanasia not appropriate, lethal trapping or shooting are the methods most commonly employed. Although lethal trapping of beavers using quick-killing, body-gripping traps such as the Conibear 330 is considered humane (Fur Institute of Canada 2014), for various reasons this method appears to be seldom employed throughout most of Europe, and would be illegal in Britain. This trap is not target species specific, and the human health and safety risk in more populated areas, especially for children, mean this is a method unlikely to be legally implemented. In contrast, beaver-hunting for recreation and to control damage is well established, particularly in Norway, Sweden and Finland, where the sale of beaver-hunting rights to both residents and visiting hunters provides income to landowners as well.

Details of hunting practice, such as the shot size for shooting beavers, are most likely to be licensing issues, and therefore determined by the SNCOs. Small-diameter shotgun shot ('birdshot', i.e. graded smaller than no. 6) is not recommended as a reasonable or humane method for killing beavers unless it is used at point-blank range, firing from behind the ear and aiming towards the opposite eye. Using a shotgun from a distance >20 m is not recommended, given the particularly thick skin and fur of beavers, as this may result in an individual which has incurred significant but non-lethal injuries entering the water as a natural defensive reaction and then drowning or dying at a later stage from its wounds or infections. For close-range dispatch, a .22 rim-fired rifle using long rifle ammunition (preferably high velocity) can also be used by experienced operators, with the rifle shot administered from above, aimed between the shoulder blades in order to destroy the heart and associated vital structures (spine, major vessels and airways). Head shots with rifles should be avoided, as the skull bones of a beaver are thick and the brain cavity small. It should be noted that if a rifle is used to dispatch a beaver, the shooter must have a firearms certificate (FAC) in the UK which specifies the species for which the firearm held can be used, so the holder would need to get 'Eurasian beaver' added to their FAC to be able to legally shoot beavers.

Euthanasia can also be achieved through humane injection by a qualified veterinary surgeon, administering sodium pentobarbitone intravenously via the ventral tail or the cephalic or saphenous vein.

6.4.5   Animal-welfare considerations

Anyone taking a beaver into captivity in Britain becomes responsible for its welfare under the Animal Welfare Act 2006. This legislation covers the release of any individuals unfit to be returned to the wild (e.g. sick or very young individuals), as wells as the husbandry practices and welfare standards to be maintained whilst in captivity (Campbell-Palmer and Rosell 2013). Suitable release-point selection should include: beaver presence (as placing an unrelated individual within an active territory could result in stress and injury), assessment of food provisions, landowner permission to reduce conflict, and likelihood of culling.

Mortalities and significant injuries have all been recorded when beaver-trapping using live snares or Hancock and Bailey traps (Davis 1984; Smith and Peterson 1988), in which beavers may be held partially in water, exposed to predators and to the elements. Welfare concerns using these methods include discomfort (Grasse and Putnam 1950), entanglement (e.g. injury, respiratory arrest: Davis 1984), stress, injury or killing by predators (McKinstry and Anderson 2002), drowning (Buech 1983; Davis 1984), and exposure to the elements (hypothermia: Grasse and Putnam 1950). Young beavers (4–7 months) lose heat faster in water temperatures of 1–12°C compared to those >1 year (Macarthur and Dyck 1990), and therefore death from hypothermia could result if these animals are held in a partly submerged trap for 4–5 h (Rosell and Kvinlaug 1998). Although Bavarian (or other box-type traps) are difficult to manoeuvre in the field, they enable trapped beavers to move, groom and feed, and offer more protection from the elements (Koenen et al. 2005). Placement of traps should avoid temperature extremes, overexposure and sudden or unexpected water-level rises. Long exposure to the sun should be avoided in beavers (Hill 1982). Pre- and post-trapping monitoring is required to ensure that no dependent offspring remain. A trapper is responsible for an individual beaver under the welfare legislation (Animal Health and Welfare (Scotland) Act 2006), as well as complying with transport legislation, i.e. the Welfare of Animals (Transport) Order 2006, which governs the suitability of transportation crates, including dimensions and ventilation, along with grouping of animals, etc. Incidental trapping of other species, particularly pregnant mammals, should be avoided through appropriate precautions including trap placement, bait, trap type and timing of trapping effort.

Bailey and Hancock traps can be triggered by a range of species including ducks, dogs, pike, otter and deer, for example, which may potentially result in injuries; and mortality has been recorded (grey heron, *Ardea cinerea*) (Rosell and Kvinlaug 1998). The health and safety of people setting such traps is an additional consideration. Inappropriate equipment, not designed for beaver use, has caused injuries to beavers including facial cuts and damage to claws and teeth. Deep layers of straw will reduce injury and slippage in transport, while apples should be provided for moisture. Beavers being translocated should be screened for significant diseases (see section 3.7 above); and, as this may take a few days to report, beavers may need to be held in appropriate captivity for a brief period. Transporting any mammalian species during its last 10% of gestation or with significant wounds is prohibited under the Welfare of Animals (Transport) Order 2006 council regulation (EC) 1/2005, unless approved by a veterinary surgeon on site. An additional consideration would be the welfare of any released individuals, particularly in relation to the identification of suitable release points.

Additional welfare considerations for any beaver undergoing veterinary procedures under anaesthetic would include the suitability of holding facilities for surgery and animal recovery, in compliance with animal-welfare legislation in Britain. The anaesthetic regime employed will impact on release times – particularly relevant to animals being released into the wild. Purely gaseous anaesthesia reduces this risk, while injectables, even used with partial reversal agents, have longer recovery times. Individual recovery time varies, so animals should be fully recovered before release near water to avoid any risk of drowning. Any anaesthesia carries some risk of fatalities. Heat loss during anaesthesia should be monitored and prevented through the use of heat pads and/or wrapping the feet and tail in tinfoil. Open-surgery sterilisation techniques are likely to result in larger areas of shaved fur, larger incisions and consequent stitching, which increase the risk of post-operative wound breakdown, infection and loss of heat (hypothermia). Consideration should be given to thermoregulatory impacts of a larger shaved area, aiming to reduce this area and/or undertaking procedures in warmer seasons. Wounds may present risk of infections. Beavers can be sterilised at any age after being weaned (3 months and above), though consideration of the proportion of shaved area and the time kept away from the rest of the family is important for kits (Pizzi *et al.* in preparation).

Any hunting would require a closed season on animal-welfare grounds to protect dependent young during the spring and summer months. The shooting of heavily pregnant females, while not considered a welfare issue if performed correctly, may be unpalatable to hunters and the general public. Ideally, any trapped animal should be humanely dispatched at the trapping site and not transported. The appropriate dispatch of animals should not present a welfare issue (Animal Welfare Act 2006).

Key concepts

- Once reintroduced to a river system, beaver populations grow and expand, and over time (e.g. 20–40 years) will occupy all accessible suitable habitat.
- As the population expands, beavers commonly start to occupy habitats which require them to modify their surroundings and this can increase the number of conflicts with humans.
- The extent and impact of beavers on human land-use is largely determined by the character of the landscape, with flatter flood plains often providing more challenging management situations.
- Effective beaver-management techniques are well developed across Europe and North America.
- Legislation associated with permitted beaver management in Britain has yet to be established, any practitioner must seek advice from the relevant statutory authorities in advance.
- The establishment of buffer zones around freshwater bodies is effective at decreasing conflicts and has wider environmental, economic and social benefits.
- Damming and burrowing are often the most common causes of conflict and can be challenging to manage, especially in artificial drainage systems and flood banks. Dam removal can be repetitive and resource intensive. Flow devices can reduce this but involve some tolerance of water-level rises. Anti-burrowing techniques can be expensive where buffer-zone creation is not possible.
- Foraging activities such as tree-felling and crop-raiding are more easily managed using different types of tree guard, fencing and other deterrents.
- Trapping and translocation are viable management tools in the medium term but are a limited solution. Animal-welfare issues, including trapping method and timing, and suitability of the release area are all vital to plan in advance.
- Culling beavers may be an option in certain situations, but beaver hunting as a management tool is unlikely to occur in Britain.
- Culling beavers is effective if it targets problem individuals – but, in this highly territorial species, it will create vacuums which are then filled with dispersers as opposed to a stable family unit.
- The culling of adult females while dependent young are potentially present is an important welfare consideration. Therefore a closed season and humane dispatch techniques are recommended.

7. Survey and monitoring

Beaver surveys will vary according to their objectives, resources and employed methodologies. While a simple survey might be appropriate at a local level to determine whether beavers are present or absent, more extensive studies will be required to determine their status on a landscape scale. Repeating standard surveys at regular intervals can identify changes in abundance, distribution and the effects of any applied management strategies. While regular visual observations of beavers can be undertaken on bodies of open water, their various field signs are uniquely distinct. With experience, their freshness can be readily assessed.

More detailed population or individual surveys can also be undertaken using live capture traps. The trapping and handling of beavers requires specialist knowledge, training and equipment (Campbell-Palmer and Rosell 2013). Trapping in Britain currently must be undertaken with the prior permission of the landowners and cannot proceed without an appropriate licence from the relevant Statutory Nature Conservation Organisations. Live trapping affords the opportunity for mark-and-release monitoring, as well as sample collection for health or genetic studies (Campbell-Palmer et al. 2014). Individual tags, biologgers or biotelemetry devices can be attached to study beaver movements and to assess territorial sizes, habitat use, and behavioural and social interactions (e.g. Campbell et al. 2005; Graf et al. 2015).

7.1 Non-invasive monitoring techniques

Several non-invasive techniques can be used to monitor beaver presence. These include carrying out a simple questionnaire survey of landowners and stakeholders in the area. Statutory organisations may also hold publicly accessible records. Aerial surveys using small planes or helicopters have been used to identify potential areas of beaver activity. For example, SNH undertook an aerial survey within the River Tay and River Earn catchments to identify potential beaver-occupied areas which were then subject to further detailed surveying on foot and by canoe (Campbell et al. 2012a). The use of drones is also becoming increasingly common. These surveys have revealed that it is possible to identify areas of beaver-felled trees from the air, while other large structures, such as ponds and dams, are also likely to be seen, though lodges may prove more difficult to identify from the air and require further foot- or water-based surveying, especially at times of year with high vegetation cover.

As beavers leave clear physical signs of their presence, the most straightforward way to survey is to look for field signs (see Appendix A) of beaver activity on foot or by canoe.

Direct visual observation or using camera traps may be useful to confirm the presence of beavers as well to enable behavioural data to be collected (Rosell *et al.* 2006).

7.2 Habitat suitability/habitat-use survey

Consulting Ordnance Survey or electronic maps is an essential component for planning any survey. As more information becomes available regarding beavers' distribution, mapping data from previous surveys can be utilised to complement larger-scale projects. Beaver surveys may be appropriate in any water's-edge environment where vegetation is readily available. While most can be undertaken on foot along navigable watercourses, surveys may be carried out from an open canoe or small boat. An important consideration for the planners of surveys is an assessment of hazards and risks, and it is advisable to complete an appropriate risk assessment (see Appendix C for an example). The basic field equipment needed to carry out a survey for field signs of beaver activity should include:

- Hand-held GPS
- Camera
- Maps
- Notebook/field sheet/field-sign mapping device
- Guide to field signs (see Appendix A)
- Binoculars
- First-aid kit and mobile phone

Although surveys can be carried out at any time of the year, beaver field signs are commonly obscured by vegetation in the summer months. The beavers' seasonal preference at this time for vascular plants rather than more obvious woody vegetation also makes evidence of their presence harder to detect. Snow may conceal field-signs and beaver activity can be reduced in winter. For these reasons, field-sign surveys are best undertaken in early spring or autumn.

While most signs of beaver activity are generally found within 10–20 m from the water's edge, some foraging trails or canals may lead inland away from the main water body (distances of 30–40 m are not uncommon, depending on habitat). These features should be followed, and often lead to clusters of further field-sign activity. Surveys are best undertaken in a systematic fashion by following the edge of waterways looking for signs, and investigating any subsidiary trails leading out into the wider environment. For a variety of reasons, some stretches of watercourse may be inaccessible, and their location should be identified for future reference. Photographs and descriptions of beaver field signs are shown in Appendix A.

When a sign of beaver activity is discovered, the following information could be recorded:

- Date
- Type of field sign/activity
- GPS coordinates
- Photograph/sketch
- Age of sign (a subjective assessment of whether it is old or recent)
- Area of sign and distance from water
- Width of river/stream and approximate depth (metres)

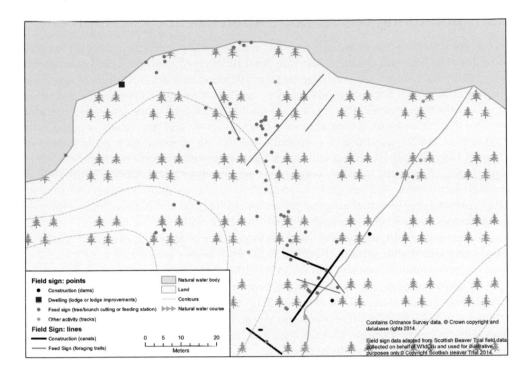

Figure 7.1 GIS mapping of beaver field signs to monitor habitat use and activity type. (K. Hill)

By plotting field signs, habitat-use maps can be created which may show seasonal variation and distribution spread over time. The type and abundance of vegetation can provide an indication of food availability. By looking at habitat use in relation to topography and nearby land-use practices, current and potential beaver impacts on the physical environment may be assessed. The River Habitat Survey (http://www.riverhabitatsurvey.org/rhs-doc/the-survey/) provides a standardised method for collecting this type of information along a set length of river of 500 m. In-stream and margin features are recorded, including channel dimensions, sediment size, river bed and bank form, features and vegetation content and morphology. Specific features of relevance to beavers include the extent of trees, shrubs and dead wood (Environment Agency 2003).

Particular field signs, such as the presence of scent mounds along territorial boundaries can be a useful clue to identify different territories and therefore family groups (Rosell and Nolet 1997; Rosell et al. 1998). In addition, lodges are robust features which remain visible in the environment long after they have been abandoned by beavers. As such, they should not be presumed always to indicate current beaver presence without evidence of fresh activity. It should be noted that beaver families often have more than one lodge in their territory. During autumn and winter, the presence of a food cache, the top of which is often visible from the bank, is a good indication that the lodge is currently in use and should be observed for new kits the following summer. These food caches are located directly in front if an active lodge or burrow.

7.3   Monitoring beaver population size and development

Precise counts of beavers for monitoring and management purposes are difficult to obtain (Rosell *et al.* 2006), with the most accurate determination of family size and composition being achieved through removal trapping (Hay 1958) and/or mark–recapture observations (Busher *et al.* 1983). Both methods require significant resource investment and effort. Signs of previous beaver activity remain visible for many years, and it is therefore possible during an autumn census to map both presently and previously occupied sites (Parker *et al.* 2002a), leading to a potential overestimate of abundance. If the location of a beaver lodge is known, sitting and watching quietly with binoculars from the opposite bank is one of the most effective ways of seeing beavers. Observation of beaver lodges to count family size and assign individuals to age classes (bank counts) can produce data representative of actual lodge occupancies (McTaggart and Nelson 2003). However, many factors, including the weather, the number of watches, the time and season at which observations are undertaken, and the experience of observer, can affect results, so should be viewed with caution (Rosell *et al.* 2006). It is very rare for all family members to be observed simultaneously, and it can be difficult to count them in sequence as they leave or enter the lodge. As a management technique for large areas, e.g. catchment scale, bank counts are normally too time consuming, require multiple observations and often may not count every individual. Bank counts can be much more useful for smaller areas, especially for beaver families that are regularly observed, over a number of seasons, e.g. during scientific trials or on nature reserves.

In larger, more open and stiller water bodies, beavers may be relatively easy to observe and will largely ignore viewers who are quiet and still. Observations are more difficult in small, narrow stream systems, marshes or densely vegetated ponds. Generally, viewing beavers is best done about an hour before dusk, continuing until the light fades, or about an hour before sunrise in an effort to see returning individuals (Rosell *et al.* 2006). Pregnant or lactating females can be identified from the presence of protuberant nipples between their front legs. Although with experience observers may be able to determine the age-classes of beavers, the identification of individuals is often impossible unless they have obvious deformities such as tail injuries, or on rarer occasions a difference in coat markings. Variance in visible body length and width of head can help to distinguish the ages of individuals in a family. Adults have broad heads; and, while swimming, their backs are generally not visible above the water line, unless they are floating. Kits (<1 year) appear higher in the water and are quite buoyant. Their backs are completely visible above the water line, and their fur seems to be almost dry. Beavers over the age of ~3 years are usually very similar in size to adults. When out of the water, feeding along a shoreline for example, any differences in size become much easier to see. Although familiarity gained from regular observations will increase the ability to recognise any individual differences, these tend to be slight and can be hard to observe. One benefit of watching an active lodge at emergence times is that it may be possible to count the occupants as they come and go (being careful not to double-count returning animals!). The reproductive status of the family can also be determined, as more than two animals may indicate either the presence of offspring or an active breeding pair. Beavers may dive and stay submerged for some time; and, when a beaver is disturbed, it may be possible to hear the characteristic slap of its tail on the water surface that precedes a dive. Their distinctive gnawing on woody vegetation can be heard on still evenings over a considerable distance. Spotlighting beavers from boats at night is also a common method

Figure 7.2 (Left) Camera trap placed on freshly used feeding station, with metal pole marked by reflective tape at 10 cm intervals for comparative purposes to distinguish size differences between beaver family members. (Right) Camera traps can monitor trap (unset) use by beavers. (Scottish Beaver Trial)

of observation to which they will become habituated when the process is repeated over time.

Camera traps are a very popular aid for wildlife studies. There is a broad range of information available on specifications of different models and designs. Camera traps are particularly useful in studies of beavers to confirm presence, e.g. occupancy of a lodge. They can be the most effective tool for recording more secretive beaver behaviours such as scent-marking, canal-excavation, or lodge- and dam-maintenance. Cameras have been used to monitor beaver-trap use, determine the presence of kits, assess the number of individuals present at a specific location, identify their age class, identify pregnant/lactating females, and monitor behaviours. Permission from the landowner should always be obtained before cameras are placed at a specific location. Their positioning should be carefully considered to ensure that images of people are avoided and to reduce the chances of theft or damage. Fresh feeding stations along the shoreline, or obvious foraging trails, are likely to afford the best results, as these features may be used by more than one individual on a regular basis. During the autumn, beavers tend to be more active in and around their lodges, engaging in lodge-maintenance and food-caching behaviours.

7.4 Distribution mapping and population estimates

The mapping of field signs can be relatively straightforward; however, to get an accurate indication of distribution, consideration should be given to the area and scale of interest (distribution within a particular area, within a catchment, etc.), and over what timescale (snapshot in time, or how/if a population is spreading). These factors, along with resources available, are likely to determine how distribution mapping will be used and what it can tell you. Distribution mapping, if not repeated, may be misleading in some areas, particularly if trying to determine whether beavers are colonising new areas or simply moving through an area leaving signs of activity but may not be resident. Repeated surveys over time, and determining where clusters of field signs are of mixed age, may indicate active territories as opposed to areas through which dispersers have passed.

Figure 7.3 Scent-marking along the shoreline at the end of a well-used forage. (R. Campbell-Palmer)

One way to obtain a more accurate assessment of distribution and population size is to estimate beaver abundance by attempting to identify the number of family groups along a stretch of river (e.g. Campbell *et al.* 2012a). In a recently colonised area, these groups may be quite well separated; but, as the population becomes established, gaps may then become occupied, with scent mounds providing indications of shared territorial boundaries between groups (see below). As beavers are highly territorial, their density is self-limiting. The number of family groups along a stretch of river will also depend on the quality of the habitat, particularly the amount of food available. Signs of beaver activity will tend to cluster around the home lodge or burrow, although a complicating factor is where a family group uses more than one lodge or burrow.

Undertaking repeated, successive lodge observations and recording individual animal counts (at dusk and dawn), after the kit-emergence period (July–September), can provide important information on the number of individuals occupying a lodge, and their age-classes (Rosell *et al.* 2006). Scaling up from family groups to a number of

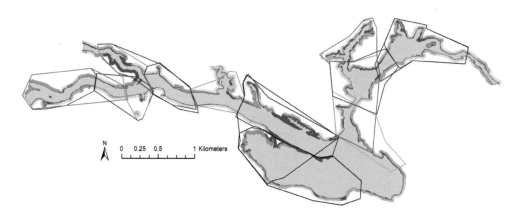

Figure 7.4 Mapping of different beaver family territories over several years through a combination of animal observation, trapping, radiotracking and location of scent mounds, River Lunde, Norway. (Campbell 2010)

individuals is usually done by assuming an average number of individuals in a family group (e.g. adult male and female, and 1–3 offspring). For example, the size of an average beaver family, from a review of 13 beaver studies in Europe and observations of beavers living at high densities in Norway, is estimated at 3.8 ± 1.0 individuals (Rosell and Parker 1995; Rosell *et al.* 2006; Parker and Rosell 2012).

7.5   Habitat assessment prior to beaver release

Where further reintroductions of beavers are planned in Britain, habitat assessments of potential release sites will be required to assess the suitability of these locations. Methods for identifying the suitability of habitat for beavers have been published by Allen (1983), with modified methods in Halley *et al.* (2009), and by Macdonald *et al.* (1997). Full protocols are currently available on the SNH website; the IUCN (2013) and Scottish National Species Reintroduction Forum guidelines (2014) should also be considered.

The main features to consider in any site assessment for beavers are:

- The composition and structure of the vegetation within 30 m of the water's edge
- The distribution and abundance of palatable riparian trees
- The character of the riparian edge habitat
- The available range of emergent or submerged aquatic plants
- The hydrology of the water bodies available to the beavers, including flow speeds, level stability and shoreline features
- Topography – gradient of land, substrate type, valley shape
- Associated land-use – disturbance, infrastructure, water use

However, it should be noted that Habitat Suitability Index models are good for species with specific needs, such as capercaillie (*Tetrao urogallus*), but tend to be much less suited for beavers which can adapt the habitat to their needs. Previously some of these models have been applied to areas and deemed them unsuitable for beavers, and which are now supporting healthy beaver populations – hence, caution and common sense should also be applied. Map based models need to take into consideration the spatial scale at which beavers select habitat. For example, riparian woodland less than 10 m deep may provide sufficient woody resources for a beaver family but is unlikely to mapped at anything other than fine scale. In time, and with growing population densities, beavers can occupy even the most modified landscapes if no other options are available.

Key concepts
- Beaver monitoring and surveying requires ecological knowledge, particularly to locate all the resting sites, to age the field signs and to determine the number of animals present.
- Trapping and handling requires some specialised knowledge, training and equipment. This is likely to require a licence; and landowner permission should be sought in advance.

- Field signs are unique and obvious, though population estimates based on these are difficult due to persistence in the environment, seasonal use/patchy feeding, population densities, habitat quality and multiple resting sites.
- Beaver scent mounds mark territorial boundaries; these mounds have a distinctive smell which is easily recognisable once learnt.
- Beaver observations are possible at dusk and dawn; look for water movement and sounds (gnawing).
- Camera traps are very useful to identify habitat use, different individuals, the presence of pregnant females/kits and secretive behaviours.
- Population size for an area is typically estimated during winter when active lodges/burrows can be identified through the presence of a food cache. Each active territory is then multiplied by 3.8 (representing the average number of family members) to obtain the population size.

8. Learning to live with beavers

8.1 Future management recommendations

It is likely that any future programme of beaver management in Britain will be an evolving process as established populations expand and applications for further releases are considered and potentially approved. This is subject, of course, to the decisions of the English, Welsh and Scottish governments regarding the future of beavers taking place at the time of writing. Any decision that allows beaver populations to remain will also include clarification on which beaver-management techniques would be permitted and under what circumstances. Any future management strategy should adopt a practical approach that is flexible, and open to revision as appropriate. This is likely to prove much more successful in the long term than a rigid system of structured, licensed controls. The strategy should be adapted to the social, economic and legal requirements of a British context. It should provide a broad range of management options and tools. Additionally, it should be acceptable to both landowners and to wider society (Hartman 1999). Drawing on the established experiences of the best management systems already in place in Europe will be key to attaining this goal.

In the immediate future, the current Bavarian beaver-management model offers a potential template for dealing with the mitigation issues which are most likely to arise in the arable landscapes of Tayside. Particularly in a Scottish context, an alternative long-term management goal could follow an approach more similar to the Norwegian Wildlife Act and Nature Diversity Act – to maintain beaver populations throughout their natural range at densities sufficient to maintain biodiversity, and undertake translocation or culling as necessary (with the future potential of optimising both recreational and economic opportunities) to reduce beaver–human conflicts and costs.

Any considered process of long-term beaver management is best focused where practicable on the establishment of buffer zones of native riparian vegetation along the length of freshwater features. The provision of riparian buffer zones of this type is a target under the EU Water Framework Directive due to the wider benefits – social, economic and environmental – that result. As appropriate, this habitat could be restored through planting-up (with native tree and shrub species) or natural regeneration. From the beavers' perspective, the greater the availability of riparian woodland with young regeneration of favoured forage species in the immediate vicinity of a freshwater body, the less the requirement for more distant forage. This option will not, however, be achievable where essential infrastructure (for example) is protected by raised flood banks, where heavily modified or managed water bodies are common, or where the associated land-use is too

Beaver management in Bavaria

The Bavarian system of beaver management has been developed over the last 20 years in order to provide a broad suite of potential solutions to human–beaver conflict situations. Although lethal control is one of the options in environments where a given beaver activity is considered unacceptable, other solutions, such as electric fencing, tree protection, dam drainage, land purchase, compensation and previously translocation, have all been applied. Where possible, the management system involves a process of forward planning to identify, when practicable, where conflicts are likely to arise. The system is funded by the Bavarian state government and the BUND Naturschutz in Bayern e.V., a nature conservation NGO. It operates via the local county (Landkreis) nature conservation agencies, which are legally responsible for beavers, supported by a team of 400 trained beaver wardens and two employed (part-time) beaver-managers. These individuals are distributed throughout the state with access to live-capture traps, holding cages, electric fences, tree protection material and other appropriate equipment. The wardens are contactable through a register maintained by the state conservation authorities when beaver conflict situations or simple questions arise within a landscape. The wardens nearest to the issue respond swiftly to assess the character and extent of the problem. In the event that it can be readily resolved, they will take any necessary action required to do so under licence from the local county administration. If the problem is more complicated, they will contact one of the beaver-managers, who will visit the site. The whole process is based on a sound knowledge of beaver ecology and of what is reasonably achievable, and on sympathetic liaison with site-managers or private landowners in order to establish their requirements and any appropriate legal constraints. This system of management has now been adopted by other German states, and the state nature-conservation bodies in other countries such as Austria and Switzerland.

commercially important for the establishment of buffer zones. Geospatial analyses are being employed in Scotland to provide an overall context of potential beaver habitat and connectivity; their likely overlap with habitats and species of conservation significance, and the identification of riverine sections less likely to be dammed by beavers (SNH 2015; Stringer *et al.* 2015). Along with population modelling, these can be useful tools in estimating population growth and colonisation, for local management planning and in wider national strategic planning (e.g. identifying areas where mitigation should be prioritised: SNH 2015).

Government 'positive payment' schemes in a number of European countries support landowners when reintroduced species are found on their land. These schemes have proven quite successful, especially in enabling the establishment of a reintroduced species or the protection of small/threatened populations, and are generally popular with the public (Cope *et al.* 2003; Naughton-Treves *et al.* 2003). Opponents of such schemes question the long-term cost implications and the ethics of potentially discouraging landowners from adapting their activities to a more wildlife-friendly approach without financial motivation (Bulte and Rondeau 2005). While there is potential for the SNCOs in partnership with the water-management authorities to incorporate beaver habitats in the landscape and biodiversity targets of the Water Framework Directive, further opportunities could be explored to support beaver-created environments where they credibly contribute to the sustainable provision of flood attenuation. In established programmes of 'Upstream Thinking' landscape-scale-based projects which seek to slow flows and maintain greater reserves of water in the higher reaches of river catchments for flood prevention, or for water purification or retention purposes, beaver-generated landscapes would be a compatible and sustainable prospect (Gow and Elliott 2014).

The opinions of the public and of landowners are critical to the success of any beaver-management programme. Most beaver-reintroduction projects (Europe and North America) report initial majority public support, though generally face opposition from landowners. The acceptability of lethal control increases over time as beaver populations expand and the reality of living again with the impacts of this species becomes an actual experience (Jonker *et al.* 2006; Siemer *et al.* 2013). Lethal control is still considered to be unacceptable in some communities that have experienced significant beaver impacts (Needham and Morzillo 2011; Perryman 2013). The culling of problem beavers is likely to be culturally more acceptable in some parts of Britain than in others.

It is important that, where undesirable impacts result from beaver activity, these are dealt with promptly and competently. An informed and well-resourced advisory system for land-managers and landowners will be a vital consideration as beaver populations expand.

Ideal key management requirements:

- Public education initiatives which assists landowners and wider society to understand the ecological importance of beavers for wildlife, habitats and water-management purposes.
- An information network to provide appropriate advice and resolve conflicts, with an ability to offer onsite assistance without the need for an overly complex and prolonged licensing system to manage animals appropriately and sustainably (i.e. for 'simpler' problems, the local beaver warden should have a general licence).
- Beaver management should be science-based, involving research and monitoring of the effects of beavers and development of flexible management strategies.
- A comprehensive, 'cost-benefit' based management plan for future beaver-reintroduction projects which identifies suitable locations for their re-establishment, identifies geographic areas which may potentially produce high volumes of beaver-human conflicts following re-establishment and importantly develops a credible management strategy for their long-term presence and incorporates stakeholder engagement.
- Ongoing proactive and site-specific management strategies.
- Long-term land-use planning and grant support to encourage the creation of buffer zones along freshwater networks and allow the natural regeneration of these areas, for multiple benefits including flood-alleviation, nutrient capture and ecosystem restoration.

8.2 Public relations, education and socioeconomics

One of the most important aspects of effective beaver management is the management of public perceptions, media relations and the adequate provision of factual information. Beavers have been absent from Britain for over 400 years. Although there is now no common or social knowledge of the species or its behaviour, beavers are a popular species (Macmillan *et al.* 2001; Gurnell *et al.* 2009). While the concept that their restoration

Figure 8.1 Beaver educational interpretation and events (Bavaria). (G. Schwab)

is environmentally beneficial is generally accepted by wider society, the prospect that there may be a future requirement for management systems which involve a level of lethal control is a more complex message to convey.

The development of a process of re-learning how to live with and manage beavers is an essential component of their effective restoration. Projects which provide information regarding the ecology of the beaver have been associated with increased levels of tolerance for its activities (Parker and Rosell 2012). This process could be overseen, supported and coordinated by the respective British SNCOs to provide easy access to factually correct information about beavers.

On a more individual level, the dissemination of information regarding beavers can be promoted through public lectures, school outreach programmes, mass media and nature walks. An example of this system of activities has been effectively developed by the Scottish Beaver Trial. Publications such as leaflets or booklets have been produced for interested organisations, local authorities and wider communities at targeted events. These materials can also be designed to support talks or lessons. It should also be recognised that the way in which the activities of beavers (and those of other animals) are perceived is often closely bound up with wider human–human social issues. The same beaver activities can be very differently perceived by a society, and subgroups within it, depending on the social context involved. Management should recognise and take account of this reality.

The potential for beaver-generated landscapes to provide resilience against flood or drought, and to capture carbon, has attracted greater attention. Wetland-restoration and stream ecosystem projects have successfully utilised beavers in Europe in just this capacity (Russia – Gorshkov et al. 2002; Baltic states – Balodis 1998; Sweden – Sjöberg 1998), while in North America the dams created by beavers over the simple flood-relief structures in the Bridge Creek Project have provided a system of much greater resilience. A wide body of independent research suggests that there is a potentially important economic role for beaver activity as a sustainable deliverer of ecological services (Gurnell et al. 2009). Although their effect will vary according to the topography and circumstance of individual sites, these may include:

- The sustainable provision of significant habitat improvements for biodiversity through activities such as tree-felling or coppicing, wetland creation and the development of microhabitats.
- The potential of beaver dam systems to assist in the purification of water by trapping silts, phosphates, nitrates and other chemicals. Beaver dams act as highly effective silt traps which can decrease downstream sediment loads. A series of three dams on the Sumka River in Russia trapped 4,250 tonnes of sediment – mainly from agricultural erosion – in one year, reducing the sediment load by 53% (Gorshkov 2003). In Brittany, channel capacity to purify/detoxify agricultural discharge has been calculated as being increased by up to 10 times on beaver-dammed streams (Coles 2006). The vigorous plant growth associated with beaver habitats can absorb significant levels of nitrates leaching off agricultural land. Studies in Germany suggest that the activity of a single beaver can trap 28 kg of nitrogen annually (Bräuer 2002). At current prices, the removal of nitrate from the drinking water supply costs £44/ kg. These cumulative impacts in the upper catchments of a major river system can have a significant positive impact on water quality downstream. Additionally, new research from North America suggests that beaver impoundments can be important carbon stores at a landscape scale (Wohl 2013).

There is potential for beaver-generated wetlands to retain significant volumes of water in upper river catchments. This can provide a significant landscape storage capacity in times of drought and may help to diminish flood events by slowing and broadening high water flows.

The reintroduction of a charismatic species such as the beaver, which can be watched at dusk and dawn, could provide a strong selling point and generate opportunities for overnight accommodation and hospitality spends in some places (Campbell et al. 2007). Evidence from Belgium, where a proactive marketing strategy was developed by the NGO responsible for beaver reintroduction on the basis of 'Beavers, Beer and Castles', demonstrates that this animal has popular appeal. Annually, this project generates approximately €50,000 for the guides, organisers and landowners (O. Rubbers, personal communication). The scheme has proven particularly popular with the owners of guest houses, hotels and pubs, who proactively encourage their visitors to attend in order to receive the resultant food and drink revenue.

As a result of significant television, newsprint and general media coverage, an interest in viewing beavers has already developed on a small scale within Britain. Evidence from the Scottish Beaver Trial has recorded visitors from all over the UK and Ireland, and a range of European countries (Jones and Campbell-Palmer 2014). Over 32,000 members of the public took part in SBT-held education events (>8,000 attending local events), including outreach classes and teacher training events, with an estimated additional 6,582 participants on walks made due to the presence of beavers in the area (Moran and Lewis 2014). The authors estimate that the value of these visitors was £355,000–520,000, in addition to the ecological knowledge gained through school engagement with SBT valued at £55,690, and the value of volunteering of £84,000. A survey of tourism-related businesses for the Tayside area found on the whole there was generally a positive attitude towards beaver presence, with ~26% noting increased turnover (Hamilton and Moran 2015). The same survey determined that 12% of land-managers in the area were incurring quantifiable costs due to beaver presence. The socioeconomic costs and benefits of beavers in Scotland are complex, and likely to change with time and beaver colonisation. These are discussed in more detail in the 'Beavers in Scotland' report (SNH 2015).

Wildlife tourism is a rapidly growing sector of the UK economy. This is recognised by government, tourism and economic development agencies. Guided walks to see beaver habitat and field signs, or to try to see the animals themselves, have proved popular in Scotland through public events held by the Scottish Beaver Trial and through more informal tours undertaken on Tayside. Landowners with the appropriate facilities may be able to offer hide-viewing opportunities if beavers are resident. Any estimate of the economic value of the presence of beavers should also take into account local amenities that may benefit, such as businesses that provide accommodation and food, and the potential for training local people as guides. For example, the reintroduction of the sea eagle (*Haliaeetus albicilla*) by the RSPB on the Isle of Mull has been estimated to generate £1.69 million per annum for the local economy (Dickie *et al.* 2006).

For those who will see no direct benefit from tourism, linked payments for water storage, water purification or ecological gains could potentially offer incentives. Financial support through agri-environment, water purification, flood alleviation, nature conservation and carbon-capture projects could be targeted towards landowners who accommodate beavers for the wider ecological benefits this species can bring. Although these solutions will need to be combined with pragmatic management of the animals, they could afford a mechanism to develop over time a niche for beavers in more intensively used landscapes. In the long term, the establishment of uncultivated buffer corridors greater than 20 m in width alongside watercourses would significantly diminish the prospect of beaver-generated conflicts. Notwithstanding the above arguments regarding the benefits of beavers, it is critical that those involved in beaver management are clear in their communications with stakeholders and the public; it should be made clear that this species is one that comes with costs, and these should be conveyed in an open and transparent manner when seeking to find sustainable management systems for humans and beavers living alongside one another. However, overall there is an ingrained issue of the unequal distribution of benefits/potential income vs. costs – some have to pay directly, while others directly benefit.

8.3 Conclusion

Our relationship with beavers has been historically and socially complex. In many parts of their current North American range, in landscapes which are more extensive and less influenced by human activity, the management of beaver populations is commonly based on a local or national process of some form of lethal control. In Western European and North American societies where public support for the lethal control of wildlife has diminished, the issue of beaver culling has become politically sensitive.

Beavers are commonly accorded the title of 'wetland engineers' with little real consideration of the profound significance of this term. There is now little doubt that many of the habitat-maintenance tasks undertaken by the human managers of riparian environments to promote biodiversity or reinstate 'natural' systems of resilience to flood or drought are mimicking the lost activities of beavers, such as willow-coppicing, the cutting/removal of semi-emergent plant species, the insertion of brash bundles in rivers to provide refugia for fish fry, canal-excavation in reed beds, or open-water creation. There is a growing political awareness that, rather than continuing to strive to develop these environments artificially, the reinstatement of sustainable natural processes via 'nature-based solutions' would be more effective (Pitt 2008).

While the character of any future relationship between humans and beavers in Britain will be dependent on a much wider social understanding of the value of these animals, it will inevitably entail a tolerance of their activities, combined with a pragmatic acceptance of a requirement for their management when their presence is inappropriate. There is clear evidence that, although opposition to beavers in Britain exists, there is a much wider body of support for the restoration of this species.

It is the authors' hope that this handbook will help provide those faced with the opportunities and challenges posed by a beaver population growing over time with the information they will require to make informed management decisions. Ultimately, successful beaver management should be an evolving process, subject to review and

Figure 8.2 Beaver educational engagement. (G. Schwab)

informed by science. The pragmatic Bavarian system of beaver management, which encompasses a broad approach of educational provision, non-lethal mitigation options and the targeted culling of animals where no other options remain, offers a model for Britain. It is fundamentally based on a desire within wider society to redevelop a relationship with the beaver that incorporates understanding and tolerance.

Key concepts

- Future management of beavers should be practical and adaptive; this is more likely to be accepted and to be successful than any rigid, licence-based system.
- A broad range of management techniques will be required, which will vary from site to site.
- Buffer zones of natural regeneration along riparian corridors (10–20 m wide) are likely to eliminate or reduce a large majority of beaver conflicts.
- Lethal control is likely to be a future requirement, though this will vary according to site and acceptability. Managing a stable family in an area will best serve to reduce the presence of other beavers.
- Beavers are popular with the general public and have great educational and ecotourism value.
- A management plan for Britain, including educational initiatives, is identified as an essential component in conflict resolution.
- Beaver-restoration has an important role in British wetland-restoration, and the potential for flood-alleviation should be further investigated.

Appendix A
Eurasian beaver field signs

A.1 Teeth marks

The teeth pattern left in beaver-felled wood is highly distinctive (Figures A.1 and A.2). When felling tree trunks or removing side branches, the patterns produced are those in Figure A.1 – a cutting action with no consumption – while the teeth marks in Figure A.2 represent bark-stripping, in which bark is consumed.

 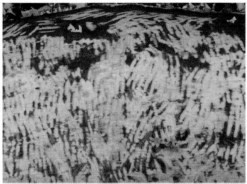

Figure A.1 Distinctive teeth marks left by beavers when felling trunks/removing branches. (R. Campbell-Palmer)

Figure A.2 Distinctive teeth marks left by beavers when stripping bark for consumption. (R. Campbell-Palmer)

A.2 Felled and gnawed trees

On saplings and smaller side branches, the angle of the cut results in a 'whistle'-shaped profile, while on larger tree trunks the beavers' gnawing and felling can result in very distinctive 'pencil-sharpened' points. Marks left by their incisors can be clearly seen or felt by running a finger across the cut end. Beavers fell timber by cutting chips from the main stem with the upper and lower incisors, alternating from one side of the head to the other, creating a distinctive scalloped pattern in the trunk. When beavers gnaw through substantial timber, they produce distinctive chippings (Figure A.3).

Figure A.3 Freshly felled tree with distinctive chippings. (Scottish Beaver Trial)

Gnawed wood, peeled sticks and wood chippings can all be classed as fresh or old by their colourations, which are very pale when fresh and darken with age (Figures A.4 and A.5).

A.3 Ring-barking/ bark stripping

Ring-barking can typically be seen on conifers and non-typical food-tree species such as beech (*Fagus sylvatica*). This is when sections of bark are pulled of the trunk by the beaver, usually in an upward motion, in thin strips. If this results in a full ring barking of a tree,

Figure A.4 Old ring-barking by beavers. (R. Campbell-Palmer)

Figure A.5 Fresh ring-barking by beavers. (Gerhard Schwab)

it is probably created as the animals walks around the trunk removing bark at the most accessible level, rather than an actual attempt to fell a tree. Ring-barking is generally unpopular, especially on large trees in public areas.

A.4 Grazed lawns and cut vascular plants

Beavers feed on a wide variety of vascular plant species and woody shrubs. On more rigid cut stems, a typical 45° angle cut is visible. Feeding by beavers on root or cereal crops is distinctive. They will completely clear a discrete area of up to several square metres

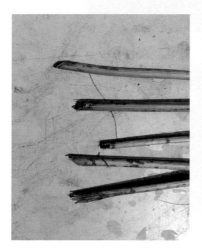 

Figure A.6 Beaver-cut bracken (*Pteridium aquilinum*) stems (Knapdale, Scotland). (R. Needham)

Figure A.7 Beaver feeding on roots of vascular plants (River Earn, Scotland). (R. Campbell-Palmer)

where the crop is adjacent to a watercourse. The site will have obvious worn access trails and, for some crop types such as maize, distinctive 45° angle feeding cuts. Root crops can be completely extracted in small patches and removed for storage in nearby burrow systems. Beavers can also produce tightly mown grazing lawns. These are created by regular feeding on vascular plants within the same area and can extend over several square metres, with a definable perimeter surrounded by longer sward. These lawns can be distinguished from the more general grazing patterns of larger water fowl such as geese by an absence of feathers and guano, and worn forage trails, leading from the water's edge, are usually nearby.

A.5 Feeding stations

Beavers often repeatedly feed at the same location on the water's edge for extended periods. Many freshly peeled sticks float away on the surface of the water into the wider environment or lie submerged in and around the feeding sites. In habitats that are densely vegetated, identification of these pale twigs in the water or on land along the shoreline

Figure A.8 Feeding station, with freshness of activity defined by (a) presence of freshly peeled sticks and, potentially, by (b) freshly cut green vegetation (depending on season). (Scottish Beaver Trial)

can often be the first indication of the presence of beavers. These feeding stations can vary greatly in size from just a few sticks to well over 100 in number. Their approximate age can be established by the coloration of the peeled sticks (lighter tends to equal fresher) (Figure A.8(a)) and/or the presence of freshly cut green vegetation (Figure A.8(b)).

A.6   Foraging trails

Beavers often forage repeatedly at points not on the immediate shoreline. Regular movements to and from these create trails and pathways (Figure A.9).

Figure A.9 Typical beaver haul-out point along a shore, and foraging trails. (D. Gow and R. Campbell-Palmer)

These features are further worn by repeated dragging of branches back to the water. Following these trails inland may lead to further field signs such as cut branches and felled or gnawed trees. These routes are usually easily visible along the shoreline, especially if snowfalls occur. They make excellent locations for positioning camera traps. Foraging trails often lead from feeding stations, which are often situated in the water or near to the shore.

A.7 Lodges and burrows

Beaver lodges and burrows are the focal points of a beaver family, providing shelter from the elements and from predators. They vary greatly in shape and size depending on the surrounding habitat and number of occupants. Some dwellings can be very obvious

Figure A.10 Free-standing beaver lodge (left). (R. Campbell-Palmer)

Figure A.11 Drained beaver pond with exposed lodge with multiple entrance and exit tunnels visible (these would usually be submerged). (G. Schwab)

structures at the water's edge, or occasionally free-standing in ponds, while others consist of burrows or chambers with few or no obvious external features. Burrows will frequently be excavated under the root plates of large trees, although burrow structures do vary greatly. Some burrows have tunnels running from the living chamber into the surrounding landscape. These are commonly used for foraging purposes. Although most are less than 5 m in length, some can extend for greater distances. Mud, branches and other vegetation are the usual construction materials for lodges. These structures can contain a number of chambers. Beavers always enter and leave these structures via a submerged entrance, often obscured by a protruding structure of woody debris. Lodges can survive as features for some time after they are no longer occupied by beavers. If beavers are in residence, fresh feeding signs or foraging trails are usually evident in the immediate vicinity of a lodge. If the branches on a lodge are carrying green leaves that have not yet wilted, this is a good indicator of current occupancy, as is wet mud or the

Figure A.12 An active lodge (no food cache as yet present, cache construction occurred later) with fresh mud layer, debarked branches and branches with fresh vegetation. (H. Parker)

Figure A.13 An inactive lodge, dry in appearance, with old branches and no evidence of fresh mud or cut vegetation. (H. Parker)

presence of aquatic plants on the outer surface. Fresh building and food-caching activity mostly takes place, and is therefore most obvious, in the autumn.

To create a lodge, beavers usually first dig a burrow or excavate a cleft in banking next to the water. In rocky environments or artificially reinforced river banks, beavers may adapt any available shallow shelves. Once a construction point is selected, it is built upon or added to by piling layers of sticks, mud and vegetation on top. Where branches protrude into the walls of the chamber, they are smoothed off through gnawing. The internal floor is lined with a bed of finely split wood shreds. Living chambers are formed in two ways. If the bankside substrates are suitable, then the beavers will excavate tunnels and chambers by digging. Where the substrates are not so suitable for digging, more of a lodge structure is built outwards from the bank itself. Burrows frequently break the surface sometime after first occupation. When they do, a 'roof' is constructed of sticks and mud; this may develop into a bank-lodge, a structure starting from the water as a burrow and ending as a lodge standing on the bank a short distance from the water.

There are, therefore, three principal lodge designs: a free-standing structure in a pool (known in North America as a 'brook lodge'), a bankside lodge and a burrow lodge. The first type is established when beavers dam a small stream to create a pool which subsequently expands to surround the lodge. The lodge is then built up in response to the deepening water. The second type is a structure entirely constructed from sticks, mud, etc., including the underwater entrance. The third is the burrow lodge described above.

It is important to be able to determine if a lodge is occupied or not, and this is best judged visually in the autumn when beavers prepare for the winter period. Those lodges that will be used that winter will almost always have fresh additions of debarked branches, mud and often some kind of green fresh vegetation, along with a food cache (which may be partly visible). Alternatively, an abandoned lodge will lack these features, and look 'dry' in appearance.

Burrows may extend from the shoreline and open above ground nearby, acting as access routes from the water's edge to food sources further up the bank. Most burrows are not visible above ground, although their entrances can sometimes be observed where the water is clear or where levels are unusually low. In narrow watercourses, these burrows are most likely to be found by viewing from the opposite bank, or from a canoe in wider or deeper watercourses. The growth of tall vegetation in the summer months can easily obscure even large beaver lodges. Beavers can also create 'day rests' (lairs), which are scrapes or shallow burrows lined with shredded woody vegetation, situated close to the water's edge, where they may sleep in preference to the main lodge during the warmer

Figure A.14 Beaver-shredded bedding. (D. Gow)

Figure A.15 A 'day rest' under a tree with shredded bedding visible. (J. Coats)

summer months. Radio-assisted observations in 11 well-established beaver territories in Telemark county, Norway, found that the beaver groups each regularly use an average of seven lodges and burrows (range: six to eight) at any given time (Rosell and Campbell, unpublished).

A.8 Food caches

Beavers often collect large quantities of woody browse during the autumn months to create winter food stores. This practice varies with climate, being more common in climates with greater risks and longer durations of winter ice cover. Cut branches are secured by jamming them into the substrate, usually at a point just outside the lodge entrance. Other branches may then be woven through or piled upon. When they are not completely utilised by the beavers or dislocated by river flows, some branches, particularly willow, can begin to regrow in the spring of the following year.

Figure A.16 Beaver lodges with food caches: mid-winter. (Scottish Beaver Trial)

Figure A.17 Beaver lodges with food caches: pre-winter, with leaves still visible on cut branches. (Scottish Beaver Trial)

A.9 Dams

Beaver dams vary considerably in height, extent and character. They can initially be formed around natural features such as fallen trees or large rocks, or where channels narrow. Dams are constructed from various materials including sticks, branches, mud, stones and vegetation. Beaver dams are faced on their upstream side by ramps of mud shovelled by the beavers and sediment trapped by the dam itself (Wilsson 1971). In some watercourses, dams are only tenable in the summer months and are destroyed by the winter flows. It should be noted that the creation of dams by beavers depends on stream characteristics. Dams are usually only built where the watercourse is initially less than 0.8 m deep and less than 6 m wide (Hartman and Törnlöv 2006), and with stream inclines of less than 2%. Dam systems can exist as multiple impoundments or as individual structures. When newly created from timber, they are unvegetated. As they age, they are rapidly overgrown and, if not kept in repair by the beavers, will gradually decline in height as their organic components decay. As the area behind a beaver dam becomes less water logged, a series of plants recolonise the area and 'beaver meadows' are formed, with associated plant succession as the area dries out.

Figure A.18 Abandoned beaver territory with unmaintained dam (lodge visible in the background on the left). A drop in water level has resulted in increased emergent vegetation and the start of beaver meadow formations. (H. Parker)

A.10 Canals

Although well-worn foraging trails may eventually fill with water, beavers will also actively excavate networks of small canals (Figure A.19) in flatter habitats. These

Figure A.19 Beaver-dug canals in Norway. (D. Halley)

structures, which branch out from a principal water body, will be cleared by the beavers of sediment, detritus, fallen leaves or other vegetation on a regular basis. This excavated debris is placed on the canal side in irregular mounds. Beavers use these canals to transport browse and building materials; they also provide escape routes back to deeper water and safety. Where conditions allow, numerous canals can be created. Most are of short length, but some can extend 150 m or more from the main water body. The average width is 40–50 cm. Canals may contain a series of small dams or muddy impoundments which can be reinforced during droughts to allow them to retain water.

A.11   Scent mounds

Scent mounds are small piles of mud, vegetation or other detritus that beavers gather from the surrounding environment. These are deposited on the bankside or on a rock or islet in a water body, and are marked using castoreum and/or anal-gland secretion. Scent mounds can be hard to find unless searched for carefully; but, if recently marked, the camphor-like smell is quite distinctive.

Figure A.20 Beaver scent mounds can sometimes be difficult to spot, but the camphor-like smell will be a giveaway. (R. Campbell)

A.12   Faeces

Normally, urination and defecation occur in the water (Wilsson 1971), and therefore beaver faeces are not often seen. The fibrous nature of the faecal content ensures that breakdown is relatively swift, especially in habitats with flowing water. Pellets of faeces ~5 cm long may occasionally be seen in the vicinity of lodges or feeding stations in areas of still or slow-moving water.

A.13   Tracks

Complete beaver tracks can be hard to find, as a heavy, wet animal which drags its tail behind it can produce distorted tracks. These are further obscured when beavers drag branches behind them. Intact prints of the fore and hind paws differ greatly. The hind paws are large and long. A clear print will display the outline of the web of skin connecting the toes. It is unlikely that this could be mistaken for other native riparian

Figure A.21 Beaver faeces trapped behind a beaver dam. It is unusual to observe beaver faeces, as defecation occurs in the water and hence faeces often break apart, though they may be recovered in shallow, still water. (D. Gow)

mammal in Britain. The smaller forepaw print is also distinctive, although the outline of every toe is not always clear. Where beaver trails are well used, individual tracks become less distinct. Muddy or snowy trails, banks with sparse vegetation and areas around their lodges, dams or canals are all good places to search for tracks.

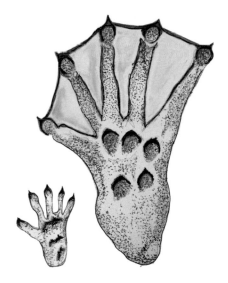

Figure A.22 Beaver tracks: forepaw and hindpaw. (Rachael Campbell-Palmer)

Figure A.23 Beaver tracks to scale. (Left) Forepaw top right. (Right) Hind (webbed) paw lower track. (R. Campbell-Palmer)

Appendix B
Diseases and parasites of the Eurasian beaver

Any health assessment of a wild or captive beaver should be carried out with specialist veterinary support and refer to current published baseline parameters for the Eurasian beaver (Goodman *et al.* 2012; Cross *et al.* 2012; Girling *et al.* 2015). Like all other rodents, beavers may harbour common European rodent pathogens (Goodman *et al.* 2012). It is recommended that any imported, wild-caught beavers are screened for the following as a minimum: *Echinococcus multilocularis*, hantavirus, tularaemia, *Yersinia* spp., leptospirosis, *Salmonella* spp., *Campylobacter* spp. and *Toxoplasma gondii* – and quarantined for rabies as appropriate.

B.1 Parasites

Beavers can host specific parasites which are of no harm to other species. These include the beaver beetle (*Platypsyllus castoris*), a small, wingless insect that lives on their skin and fur (Peck 2006). Although these beetles may be mistaken for fleas, they do not jump and are a rusty orange/brown in coloration (Figure B1). Both the larvae and adult beetles feed on the epidermal tissue of the beaver's skin without any visible signs of irritation (Wood 1965). Beetles have been recorded on Scottish-born beavers (Duff *et al.* 2013). There is only a single record of a host-switch to an otter (Belfiore 2006), which possibly occurred when the otter was present inside a beaver lodge (Peck 2006). The pupal stage of the beetle is found in the earth of beaver lodges or burrows (Peck 2006). Due to desiccation or extremes in temperature, adult beetles do not survive for long when they leave their host (Janzen 1963).

More than 45 species of *Schizocarpus* mite are associated with the Eurasian beaver (Bochkov and Saveljev 2012). These mites are usually spread through direct contact; and, although more than 10 species may live on an individual beaver at any one time, they are often restricted to a particular area of the body (Bochkov and Saveljev 2012). Beavers may harbour ticks, but these parasites do not appear to be present in all individuals or in significant numbers and are probably largely removed by the beaver's specially adapted grooming claw (Figure B2). As with other species, ticks can transfer haemoparasites (blood-borne parasites) and Lyme disease (bacteria *Borrelia* spp.) to their hosts. Although haemoparasites have not been recorded to date in Eurasian beavers (Cross *et al.* 2012), further investigations are encouraged.

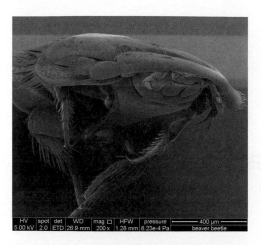

Figure B.1 The beaver beetle (*Platypsyllus castoris*), a host-specific ectoparasite of beavers. (S. Gschmeissner, R. Zange.)

Figure B.1 Grooming claw found on the webbed hindfeet of the beaver, used to keep hair 'unclumped' to maintain effective insulation and assist in removing ectoparasites. (S. Jones)

Travassosius rufus is a stomach nematode specific to beavers and has a direct life cycle, with its eggs being expelled in beaver faeces (Åhlén 2001). These thin worms can be identified within the beaver stomach walls and stomach contents at post-mortem, or by the presence of eggs in the faeces.

Stichorchis subtriquetrus is a specialised trematode or intestinal fluke found only in beavers, mainly in their caecum and large intestine. Its life cycle incorporates an aquatic snail as the intermediate host, which infects a beaver when it ingests aquatic plants (Bush and Samuel 1981). This host-specific parasite has been found in Scottish-born beavers (Campbell-Palmer *et al.* 2013).

Cryptosporidium spp. is an intestinal protozoan parasite that causes disease in the small intestine, particularly in immunocompromised or naïve individuals. It is zoonotic species that causes diarrhoea, usually self-limiting, in humans. The transmission route

Figure B.2 Beaver intestinal fluke (*Stichorchis subtriquetrus*), adult fluke (left) and adult fluke in opened caecum (right); this fluke is largely found in the caecum and large intestine. (J. del Pozo and R. Campbell-Palmer)

is faeco-oral. It is a common condition of various wildlife and domestic animals already present in the UK; beavers are most usually infected by domestic cattle.

Giardia spp. is another intestinal protozoan parasite. Although beavers have been implicated in human outbreaks – a media-inspired name being 'beaver fever' – it should be noted that this parasite naturally occurs in the environment and is present in wildlife, domestic livestock and human populations within the UK, which may provide a source of infection to the beaver (Morrison 2004). This parasite lives in the small intestine and can cause diarrhoea and abdominal pain in all mammals. In a survey of 241 wild-trapped Norwegian beavers, none tested positive for *Giardia* (Rosell *et al.* 2001).

Echinococcus multilocularis (also known as the fox tapeworm) is a pathogenic zoonotic parasitic present throughout Central Europe. There is currently no evidence of its existence in the wild in the UK, though the potential for unscreened, directly imported beavers to introduce this parasite has received attention (Simpson and Hartley 2011; Pizzi *et al.* 2012; Kosmider *et al.* 2013; Campbell-Palmer *et al.* 2014). Barlow *et al.* (2011) diagnosed *E. multilocularis* post-mortem in a captive beaver which had been wild-caught and imported from Bavaria years previously. The definitive host is usually the Red Fox, but domestic cats and dogs can also function as definitive hosts (Eckert and Deplazes 1999). Various rodents, mainly voles but rarely including beavers, function as intermediate hosts through which the tapeworm can be transmitted when they are consumed by a fox (Janovsky *et al.* 2002). Humans can also act as an accidental intermediate host, with the mean annual incidence rates ranging from 0.02 to 1.4 cases per 100,000 inhabitants in different parts of Europe; infected people often require life-long medical treatment (Eckert 1997). Captive collections should ensure that any beavers imported from areas where this parasite is endemic are not allowed to be scavenged by carnivores, as the parasite can only become a threat to humans if it gains access to a carnivore's digestive tract. As beavers are intermediate hosts, they do not shed this parasite which only completes it's life cycle in a carnivore host. Therefore beavers can not spread this parasite in their faeces (or therefore be tested for parasite eggs via faecal screening), and can not be treated for this parasite via 'worming'. Post-mortem examinations should incorporate investigations for this parasite through examination for liver cysts (Barlow *et al.* 2011). A blood test for beavers for the presence of this parasite has been developed (Gottstein *et al.* 2014); previously, screening for this parasite in live beavers involved combined ultrasound and laparoscopic investigation, particularly of the liver and abdominal organs, as was done for wild-caught beaver of unknown origin within the River Tay catchment (Campbell-Palmer *et al.* 2015).

B.2 Bacteria

Franciscella tularensis is a bacterium that can be found in beaver blood, faeces, organs and body fluids. This bacterium can be passed to humans via handling infected beavers, or water sources containing a dead infected beaver, or through biting insects such as mosquitoes and ticks which have fed on the blood of infected beavers. The resultant disease can be fatal, with septicaemia developing rapidly in some cases. It should be noted that this bacterium is not currently present in Britain, and therefore serological and or faecal culture screening is advised for imported beavers. It has been reported predominantly in North American compared to Eurasian beavers (Wobeser *et al.* 2009).

Yersinia pseudotuberculosis is a bacterium commonly found in any rodent and may be recovered from beavers (Nolet *et al.* 1997; Gaydos *et al.* 2009). It is generally carried in the gastrointestinal tract but may occasionally result in septicaemia in mammals and birds, particularly rodents, primates and some ruminants. It can also produce a chronic wasting disease with granulomatous changes in the gastrointestinal tract. *Yersinia enterocolitica* may also result in septicaemia and gastrointestinal disease and has been isolated from North American beavers (Hacking and Sileo 1974).

Salmonella spp. may be carried by any mammal and may result in zoonotic transmission to humans, causing diarrhoea and vomiting. Beavers, as mammals, may become persistent carriers and shedders of this bacterium, although it is uncommon in our collective experience (Rosell *et al.* 2001). Faecal culture is recommended.

Campylobacter spp. may also be carried by any mammal, including beavers, and may result in zoonotic transmission to humans, causing diarrhoea and vomiting, but is again uncommon (Rosell *et al.* 2001). Faecal culture is recommended.

Bacillus piliformis is a bacterium causing Tyzzer's disease, which is common in rodents in the wild and in captivity (Wobeser *et al.* 2009). It can result in acute septicaemia and death, or in a chronic wasting disease with granulomatous lesions in the liver and intestinal tract. It is passed via the faeco-oral route and may be detected in carriers or persistently infected animals in the faeces. It is uncommon in beavers.

Leptospira spp. may be carried by rodents, including beavers (Nolet *et al.* 1997). This zoonotic disease is already common in British wildlife, and can cause interstitial nephritis. It is transmitted through urine and mucocutaneous junctions. It may be tested through serological analysis using monoclonal antibody tests for servars of pools 1–6 in the first instance.

B.3 Fungi

Trichophyton mentagrophytes is mainly found in rodents, which may carry it without clinical signs. Colloquially it is referred to as 'ringworm'. Other species of *Trichophyton* and *Microsporum* spp. may also cause disease. Fur-brushing for culture in dermatophyte media is advised to assess for carrier status. This organism is widespread in the UK and Europe.

B.4   Yeasts

Adiaspiromycosis (*Emmonsia* spp.) is a yeast which has been reported in beavers and has been seen in other semi-aquatic and fossorial species, such as European Otters and Hedgehogs (*Erinaceus europaeus*) (Morner *et al.* 1999). It affects the lungs, and results in pneumonia. Beavers with respiratory disease should be screened by chest radiographs and broncho-alveolar lavage to determine the causal organism.

B.5   Viruses

Hantavirus has never been reported in beavers; but in theory, as any rodent, they may act as a reservoir. It is known to be a problem in rodents, particularly rats, and has been reported in the UK (Jameson *et al.* 2013). The infected animal persistently excretes the virus in saliva and urine. It can be zoonotic, and produce haemorrhagic fever with renal failure. In North America, a pulmonary form has also been reported. Polymerase chain reaction tests on urine and saliva may be used to screen rodents. Serology may be used to determine exposure.

Rabies virus has also not been reported in Eurasian beavers but may affect any mammal. Screening of the live animal is not currently possible, so any imported beavers to the UK should be sourced from rabies-free areas or quarantined according to current Rabies Importation Order (as amended) directions.

Appendix C
Beaver-management techniques

This section covers current best-practice recommendations for beaver-management techniques regularly employed across Europe and North America. The regulatory framework for beavers in Britain is likely to alter significantly in the future, so the advice here should be checked against current recommendations by the appropriate statutory agencies.

Additional sources of information

> http://www.beaverdeceivers.com.
> http://www.beaversww.org/
> http://www.beaversolutions.com/
> http://www.vtfishandwildlife.com/.../Best_Management_Practices_for_
>   HumanBeaver_Conflicts.pdf
> http://www.kingcounty.gov/environment/animalsAndPlants/beavers/solutions.
>   aspx
> www.biberhandbuch.de/
> www.clemson.edu/psapublishing/PAGES/AFW/AFW1.PDF
> http://www.snh.gov.uk/protecting-scotlands-nature/beavers/

It must be emphasised that in the following descriptions of management devices, competent and robust construction from good-quality materials is necessary, which will make them more cost-effective in the long term. Experience from other parts of the world has established that the construction of a poor-quality device which soon fails, leads to the erroneous conclusion that the method itself is ineffective. Seeking advice and assistance from experienced personnel is recommended.

C.1   Flow devices – dam piping

Issue   raised water levels due to beaver damming activity.

Basic principle pipe inserted through the dam and positioned to allow the water level to be lowered.

Result   control of water level behind beaver dams, or regulation of water levels at sites vulnerable to damming activity.

Materials
- 25–40 cm diameter 'soft' polyethylene (PE) plastic pipe, double-walled (more hydrodynamically effective than single-walled as they are not corrugated internally), PVP pipes may also be used, ideally 10–15 m in length (the further the inflow is from the dam, the more likely it will not be noticed by beavers).
- Pipe connectors.
- Heavy (~6.25 mm), large-mesh (<15 × 15 cm) steel netting/welded wire fabric/mesh (e.g. ASTM A185 and A497). If used (Square Fence™ filters), wooden frames may be constructed with suitable timber and associated fixtures.

Method
Pipe:
- To ensure a double-walled pipe sinks insert small holes through the ribs of the PE pipe all the way along the length to allow water in and air out. This can be done with a circular saw set to the correct depth of cut, so each raised corrugation is perforated.
- Perforate the pipe at intervals all the way through to the inside. This allows any bubbles to escape, which otherwise might accumulate and cause the pipe to rise in an inverted 'U'.

Mesh filter – this is a 3D mesh basket, traditionally known as 'filter'. This can be square or round, and is used to prevent beavers from accessing inflow:
- Bend mesh into shape, so that it forms a 3D oval shape, with mesh covering top and bottom; alternatively form a square filter which is cuboidal, and mesh supported by a wooden form.
- Bend the cut ends of the mesh back on themselves to hold the filter together.
- Cut a hole in one side, to match the diameter of the pipe used.

Installation:
- Create a breach in the dam and dig a trench down to the desired water level.
- Place the drainage pipe through the breach, with the outflow ~50 cm downstream of the dam.
- Insert the end of the pipe into the filter.
- Position the attached pipe and filter upstream of the dam.
- Secure the pipe to the stream bed at both ends and in the middle to prevent water current from altering the desired positioning.

Tips
- Pipe diameter must be sufficient to deal with the site's water flow; pipes <25 cm in diameter are not recommended (unless used in continually low-flow sites). More than one pipe can be used if required.
- Hard plastic pipes made from polyvinylchloride (PVC) are more difficult to work with in larger sizes (30 cm and above) given their weight, but are often used in Germany.
- If the water is allowed to run freely out of the pipe at a height above the ground (forming a low waterfall), cover the outlet area with a horizontal, slightly arching piece of weld mesh so beavers cannot physically stop the flow by obstruction.
- If using mesh without plastic or epoxy coating, thicker mesh wire diameter may be required, especially in acidic waters. In waters with a pH of 7 or above, thinner mesh can be used; but, due to many other stressors (ice, floods, dam weight, etc.), it is

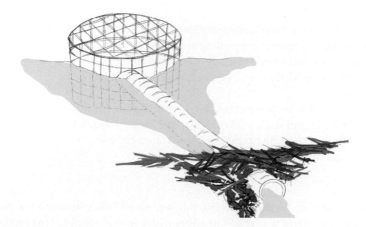

Figure C.1 Piped beaver dam (Castor Master™) with mesh filter (Round Fence™) to discourage beavers from blocking the inflow pipe. This is often submerged to mask the sound of running water. Note that braces (not depicted) should be used to hold the pipe down in place. (Rachael Campbell-Palmer)

Figure C.2 Polyethylene (PE) pipe. Insertion of holes through both walls to let any trapped air inside the pipe escape. (R. Campbell-Palmer)

advantageous to use heavier-gauge mesh. A larger mesh pattern collects less floating debris, which also contributes to the filter's effectiveness.
- Mesh-square dimensions of 15 × 15 cm excludes beavers (although kits may be able to pass through – a double layer can be created if this is an issue) while allowing most other animals to pass through.
- If properly designed, the filter should not need regular cleaning, though this should be monitored and debris cleared if required.

Figure C.3 Round Fence™ filter (left) before placement below water level (note green pontoon to manoeuvre frame before submerging). Square Fence™ filters (right) have wooden frames that can be decked over for recreational purposes, or left open in permanently shallow water where beavers have no opportunity to enter the mesh unit and plug. (R. Campbell-Palmer and S. Lisle)

Figure C.4 Flow device in an arable area where beavers have repeatedly attempted to dam. Though not visible here this flow device is also a Swept-Wing Fence™, in which fencing extends further onto the land, to discourage beavers from entering a vulnerable agricultural canal. By allowing beavers to dam along the fence line, any resultant dam will be effectively piped. Note the Round Fence™ filter to prevent the pipe from blocking at the intake. (S. Lisle)

Figure C.5 Square Fence™ filter, with pipe submerged and not visible. A roof has been fitted as the site experiences changing water fluctuations enabling beavers to enter and block pipe inflow during periods of high water flow if left unprotected. (S. Lisle)

C.2 Flow devices – culvert protection

Issue blockage of culverts by beaver activity.

Basic principle fencing constructed around a culvert to prevent beavers from accessing and blocking the culvert.

Result beavers unable to directly block culvert..

Materials
- Heavy (~6.25 mm), large-mesh (10–15 × 15 cm) steel netting/welded wire fabric/mesh (e.g. ASTM A185 and A497), wooden fence posts, screws, staples.

Method

- Place fence posts in a square with the side against the culvert open. The form can be varied to fit the contours of the stream. The sides of the fence should not touch the bank, if possible.
- Fix mesh to the fence posts. The fence should cover the area from the bottom to about 50 cm over water level.
- Build a second fence (at least 60 cm high) on the inside between the culvert inlet and the road, so that beavers cannot gain access from the land.
- To control water levels in the event of damming, a pipe and mesh filter should be inserted in addition (see Appendix C.3 above). The pipes should be placed as low as possible so that water behind the culvert is kept shallow enough for the beavers to feel too exposed to use it.

Where it is desired that the beavers (and otters and other animals) can use the culvert for passage (e.g. to reduce the chances of road accidents when beavers have to cross the road), this can be done as follows:

(a) Include a polyethylene pipe T-joint (~20 cm in diameter), placed in the fence a little under water level. The size permits passage but makes it difficult/uneconomical for beavers to pull branches or various debris through. NB: the ribs of the pipe must be double punctured prior to installing the device so that water is allowed in and pushes air to out, so that the pipe sinks. Positioning is important.

(b) To build a Turtle-Beaver Door™ in a flow device (Lisle 1996, 2003). The beavers swim in through a hole cut in the mesh (~25 × 25 cm), below water level and through a channel (~30 cm wide), and then swim out. The U-shape with two right angles prevents beavers from taking sticks or other debris through. Note that fish can easily pass through the fencing. This design may be vulnerable to blocking if beavers build a dam against the fence.

(c) Where the terrain permits, a 'beaver port' can be constructed by simply lengthening the fence against the culvert, enabling beavers to walk along a narrow path between the road embankment and the fence, and into the culvert. This requires a fairly steep road embankment, so that the beavers cannot carry sticks down the path.

Figure C.6 Flow device to protect a culvert. (S. Lisle)

Figure C.7 T-pipe mounted in a flow device (Turtle-Beaver Door™) to permit passage of beavers (and other animals) while preventing sticks and other materials from being carried in. (S. Lisle)

Figure C.8 Beaver Deceiver™ with beaver door to allow animal passage but making dragging of branches very difficult. Steel mesh is strong enough to be self-supporting, so the number of support posts is reduced. This management technique has made use of features associated with this road culvert to be effective against damming and to prevent beavers from carrying building materials. Note: the fencing mesh has been staggered in order to prevent entry of smaller beavers. (S. Lisle)

Figure C.9 Any beaver mitigation should ensure the passage of other animals; otter and turtle doors have been created using appropriate mesh size. (S. Lisle)

Figure C.10 Examples of Beaver Deceivers™ protecting culverts. (S. Lisle)

Tips

- Depending on the location it may be easier to construct the fencing on land adjacent to the site.
- Culvert fencing flow devices are often the most long-lasting and effective solution against beaver blockage of culverts, or when it is desired to prevent this from happening.
- Fitting the form to the contours of the stream is often more aesthetically pleasing.
- A robust structure of quality materials is much more cost effective not only because it lasts longer, but because beavers often build dams against the structure, it must be strong enough to bear such a load.
- The shape of the flow device can vary to fit the form of the stream and (so long as function is not compromised) for purely aesthetic considerations.
- A culvert fence combined with a pipe system represents a general strategy called a Double Filter System' (DFS).

Culvert structure and design are important factors in reducing the impact of beaver blockages and so reducing the need for management. Larger, wider culverts (~5.5 m), the width of which is equal to or greater than the watercourse, may reduce the potential for beavers to alter the flow of the water body. When installing culverts, the creation of a depression at the inlet should be avoided, as the existence of shallow pools can encourage beavers to expand these features by potentially damming.

Many of these system were originally developed in the USA by Skip Lisle (Beaver Deceivers International Inc. www.beaverdeceivers.com), it has been an evolving process that has been replicated and further developed across North America and Europe. Each case can be variable and site specific, with the basic principles of flow devices presented here. Taking time to discuss with those with previous experience can save time and resources.

C.3   Dam-removal/dam-notching

Issue   raised water levels due to damming activity/passage of fish affected.
Basic principle removal of all/part of the dam structure.
Result   water level returned to previous status/reduced flow over dam created.

Materials

- Forks, pick axe, grapple ropes

or

- Mechanical excavator.

Method

- Manually dismantle and remove the desired amount of material from the watercourse.
- Mechanically remove the desired amount of material from the watercourse.
- Leave material on bankside above high-water level (beavers will not re-use older material).

Considerations

If a dam is protecting a lodge, the time of year is a consideration in its removal. During the breeding season, females and young are vulnerable to sudden changes in water levels; and during winter, dams protecting lodges and food caches should be left *in situ* to avoid abandonment of lodges and loss of food caches.

Dam-notching and removal in Scotland can currently occur on watercourses without prior authorisation from the Scottish Environment Protection Agency (SEPA), using hand tools, rope or grapnels, providing that such work is undertaken without causing pollution, including the escape of 'silty water' downstream (SEPA 2014).

C.4   Burrow management

Issue   bank collapse/bank instability due to beaver burrowing.

Basic principle preventing access to current burrows.

Result   beavers unable to access previously excavated burrows.

Materials

- Heavy (~6.25 mm), large-mesh (~15 × 15 cm) steel netting/welded wire fabric/mesh (e.g. ASTM A185 and A497).

Method (only to be undertaken on burrows that are empty)

- Open the burrows mechanically along their course.
- Backfill with stone, or fill sandbags with a dry cement/aggregate mix and place them back up the opened burrow to a depth of ~2 m from its point of contact with the watercourse. The sandbags will readily mould themselves to the contours of the burrow and then set rapidly within 24 h.
- Cover entrance holes with wire fabric/mesh pushed down into the bed of the watercourse and extending to at least 2 m on either side of the burrow entrance.

Considerations

Extreme care must be taken to ensure burrows are empty. These techniques are only dissuasive, and it is quite probable that other burrows will be excavated elsewhere.

C.5 Bank and flood-bank wall protection

Issue   bank collapse/bank instability due to beaver burrowing.

Basic principle preventing burrowing activity with hard reinforcements.

Result   prevention of all mammal burrowing activity.

Materials

- Heavy (~6.25 mm), large-mesh (~15 × 15 cm) steel netting/welded wire fabric/mesh (e.g. ASTM A185 and A497) (4 × 4 cm mesh is used in Germany to help prevent against other burrowing animals).
- Pea shingle.
- Sheet piling.
- Rocks/stone gabions.

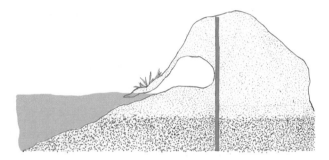

Figure C.11 Sheet-metal plates or heavy-duty wire mesh inserted into the centre of a flood bank. In relation to beavers, this has been employed on limited sections of flood banks to surround or enclose an existing lodge to prevent further burrow and lodge expansion. (Rachael Campbell-Palmer)

Figure C.12 Flood enforcement of interlocking metal sheets on the River Danube; these have the additional benefit of protecting against beaver (and other animal) burrowing. (R. Campbell-Palmer)

Figure C.13 Stone facing of river, canal or flood banks to discourage burrowing. (Rachael Campbell-Palmer)

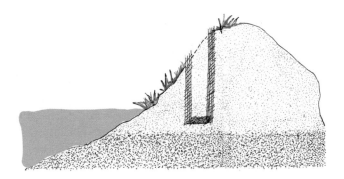

Figure C.14 Heavy-duty wire-mesh-lined trench, filled with gravel, to discourage burrowing. Note that just wire mesh or just a trench filled with gravel would work as well. (Rachael Campbell-Palmer)

Method

- Excavate narrow trenches with a mini-digger along the top of the retaining bank-head to a depth of ~0.5–1 m below the normal water level, to prevent beavers digging underneath.
- Insert sheets of wire fabric/mesh square concrete-reinforcing mesh.
- Backfill with pea shingle.
- When beavers attempt to excavate burrows, pea shingle constantly collapses into their holes, and the mesh prevents them from proceeding any further (see Figure C.14). Obvious collapses in the pea-shingle trench can then be backfilled.

or

- Insert an interlocking system of sheet-metal piling through the centre of the main flood structure to a depth that is equal to that of the watercourse.

or

- Face the river bank with large rocks or stone gabions to the bottom of the watercourse.

or

- Insert 5 cm square welded wire fabric into a narrow excavated trench 0.5–1 m below the normal water level.

Tips

- Conventional livestock netting to prevent burrowing is not advisable. This method, although cheap, is generally ineffective, especially if the underwater environment is already pitted by beaver excavations or burrows.
- While the best mesh type to use for natural contours is chainlink, the wire utilised in this mesh is soft and pliable, and beavers can rapidly open large holes in this material by pulling it with their teeth or paws. Even when it is plastic-coated, submerged mesh will corrode at different speeds, and lighter mesh gauges can be bitten through by beavers with relative ease. Mesh designs with unlocked joints can also be easily distorted by pulling.
- Other trialled techniques, such as the use of various geotextiles in new construction sites, have not proven to be effective against burrowing mammals.
- Isolating of visible lodges on the river bank with sheet piling to prevent their extension can be undertaken; but this is only temporary 'fix' for individual nuisance structures, and cannot be applied as a mitigation strategy along extensive lengths of watercourse.

C.6   Individual tree protection

C.6.1   Mesh protection

Issue   beaver browsing and felling of trees in undesirable circumstances (specimen/ornamental trees, trees by roads etc.).

Basic principle   wire mesh used to protect trees from beaver activity.

Result   beavers unable to target protected trees.

Figure C.15 Mesh tree guards to deter beaver gnawing, with expansion capabilities to allow for growth and ensure mesh is not flush against the trunk. (D. Gow and R. Campbell-Palmer)

Materials

- Light 16-gauge 0.5 in (~1.2cm) weldmesh.

or

- ~15 cm mesh in 8-gauge wire.

or

- 2.5 × 5 cm weldmesh.

Method

- Cut mesh to desired size, wide enough to wrap round the tree once and to a minimum of ~1 m height (higher in areas with heavy snowfall).
- Wrap mesh around the tree and fix at the top, middle and bottom, keeping around 15 cm away from the trunk (if flush beavers can still gnaw at the trunk)
- Cover any buttresses with mesh.
- If the nesh is not self-supporting install posts.

Tips

- Standard chicken/rabbit wire can be used at sites where the beavers have an alternative food resource; however, in general this is not recommended as if a beaver is determined to gnaw on a particular tree they have the strength to pull down this malleable material.
- Wrapping a tree in double the length required and using a system of three springs with small hooks on either end can easily deal with tree growth. As the trees expand in girth over time, the springs pop off and the guard can be replaced/expanded.
- If mesh is too light or lose, a beaver can pull it down to expose the trunk, or push underneath to access the trunk.

C.6.2   Paint protection

Issue   beaver browsing and felling of trees in undesirable circumstances (specimen/ornamental trees, trees by roads, etc.).

Basic principle abrasive paint applied to trees to protect trees from beaver activity.

Result   beavers dissuaded from targeting protected trees.

Figure C.16 Experimental site trialling Wöbra® (nearest tree and fourth tree) and sand–paint blend (second tree) vs. untreated (third tree). (R. Campbell-Palmer)

Materials

- Abrasive anti-game paint Wöbra® (https://www.fluegel-gmbh.de/media///fluegel/hochgeladen/absatz/biber-201009-engl.pdf)

or

- Exterior latex based paint
- Fine sand (0.75–1.0 mm grain size). See http://www.beaversolutions.com/tree_protection.asp.

Method

- Mix 140–225 g of sand in 1 litre of paint.
- Apply paint to height of a minimum of 90 cm (higher in areas with heavy snowfall).

Tips

- Make small batches, and stir frequently.
- Colour matching the tree bark as far as possible, can give a more natural effect.
- Exclude painting trees which are less than 2 m high to avoid damaging them.
- Check the paintwork occasionally and add a new coat as necessary (experience suggests this would be about every 5–10 years on average for Wöbra® paint, whereas the sand-paint version should be checked annually).

Considerations

This method is not always 100% effective, especially in comparison to protecting trees with wire, but may be useful in sites where wire protection is not deemed suitable.

C.7   Electric fencing

Issue   beavers accessing undesirable areas/crop foraging/raising dam height/damming above a flow device.

Basic principle temporarily deterring beavers from accessing an area/site/building dam height using electric current.

Result   beavers dissuaded from accessing the area/site/position.

Figure C.17 Electric fence arranged to deter beavers. (R. Campbell-Palmer)

Materials
- Standard electric fencing system, length as required.
- Single-strand nylon wire or electrified tape have both been used successfully.

Method
- Place 75 cm long fence poles about 1.5 m apart, depending on terrain requirements (every 5 m is sufficient in completely flat terrain; in very uneven terrain up to every 1 m).
- Place two fence wires at 15 cm and 25 cm above ground level.
- If preventing dam-building, string across the watercourse above the normal water level.

Tips
- Ensure lower fence wire is neither in contact with the ground nor over 15 cm above it at any point.
- Electric fences are ineffective where the animals can walk or swim underneath, but beavers do not normally attempt to burrow under them.
- This method has proven useful for protecting gardens and trees, and for example in hindering beavers moving from an area in which they are established to a nearby pond where beavers are unwanted.
- Beavers usually take the shortest route from water to the food source, and avoid areas lacking a direct 'emergency exit' to the water.
- It has been shown that in 90% of cases, deploying a fence for 1 week will protect crops for a further 2–3 weeks. One unit can therefore be used to protect several areas.

Considerations
Use with caution, as beavers have been known to clasp power wires with their teeth, resulting in a number of deaths.

C.8   Permanent exclusion fencing

Issue	beaver activity in an undesirable area, or retaining beavers within a captive setting.
Basic principle	installation of fencing to exclude/retain.
Result	beavers unable to access/leave an area.

Materials
- Galvanised high-tensile mesh, or similar – with a mesh dimension of no more than 10 cm.
- Locked-joint or weldmesh.

Method
- Install fences to a minimum of 120 cm above ground.
- Extend a ~40 cm skirt of mesh, and peg into the ground facing towards the side which beavers will approach.
- Bury a section of fencing vertically underground.
- Fit one single-strand electric-fence wire on the beaver side of the fence ~30 cm from the ground to prevent it from shorting out on vegetation.

Figure C.18 Permanent fencing specifically designed to retain beavers. Note hotwire fencing and mesh under-skirting (Devon). (R. Campbell-Palmer)

Tips
- To protect a grove of trees within a riparian zone, a fence parallel to the water body may be enough to deter feeding, although it should be noted that if this is the only area of suitable feeding resources in the area, beavers may be more determined to access these resources, and so more robust fencing will be required.
- It is important to ensure there are no gaps or weak points in the fence line that the beavers could exploit; and, if beavers do access such an area, investment against burrowing under fences should be made.
- Fences are most effective when they are positioned well back (~20–30 m) from the edge of a watercourse.
- Hinge-joint netting or chainlink should be avoided, as the verticals can be easily pushed out of alignment. Similarly, mild-steel netting is not desirable, as it is easily distorted and the mesh bent out of shape.
- Any electric fencing must be checked regularly to ensure it is functioning effectively.

Considerations
Use electric fencing with caution, as beavers have been known to clasp power wires with their teeth, resulting in a number of deaths.

C.9 Deterrent fencing – ditches and small streams

Building a deterrent fence of this type isolates the watercourse upstream or downstream from the rest of a beaver population in the area. Although the isolation effect is not 100%, it considerably reduces the chances that a beaver will move into the upstream section, and it considerably reduces the chance that two beavers of the opposite sex will do so at about the same time.

Issue   potential beaver dispersal into an undesirable area.

Basic principle installation of fencing to dissuade beavers from dispersing past a set point via ditches and streams.

Result beavers unable to move past a certain point.

Materials

- Strong steel mesh of about 10 × 10 cm mesh size in the stream itself.
- If movement of larger fish (e.g. large spawning salmon) is expected, 15 × 15 cm can be used
- Normal chicken-wire (or similar) fencing, preferably of good quality, for use on land.

Method

- Build a 1 m high fence (higher in areas with deep snow levels) downstream of the area to be protected, preferably in terrain that is not in itself suitable as beaver habitat (so that, if possible, no beavers are resident immediately downstream of the fence).
- Bury the mesh in the streambed to prevent beavers swimming underneath.

The inner 'sock' runs downstream on both sides from the 'wings', with a slight inward taper towards the stream (this 'guides' beavers on land back towards the stream). The minimum distance at the outlet to the 'sock' should be at least 5 m (otherwise there is a danger that beavers established just downstream of the installation may burrow from the stream under the fence). The 'wings' of the installation extend from the stream grille, on both sides of the stream, ~80°, slightly in the downstream direction. An extension at right angles, pointing downstream, is added at the ends.

This design guides any dispersing beaver moving up the stream and encountering the grille back to the water downstream, should it get out of the water at this point. To go around this device requires swimming or walking about 100 m back downstream, then walking 100 m away from the water when the beaver has no way of knowing that this would allow it to pass the grille. Moving such a distance from water, especially in unfamiliar terrain, would make the animal highly vulnerable to predation in a natural situation, and as such is instinctively avoided.

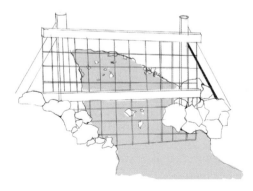 

Figure C.19 Deterrent fencing across a small stream at the Scottish Beaver Trial, note weld mesh used across water and partially sunk into stream bed. Size selected so as not to impede fish passage, with standard stock fencing used for terrestrial components. (Rachael Campbell-Palmer, Scottish Beaver Trial)

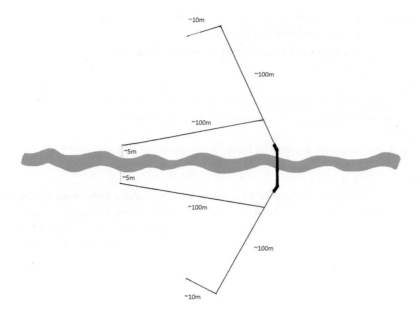

Figure C.20 Fence on stream/ditch to hinder beaver dispersal upstream. (Rachael Campbell-Palmer)

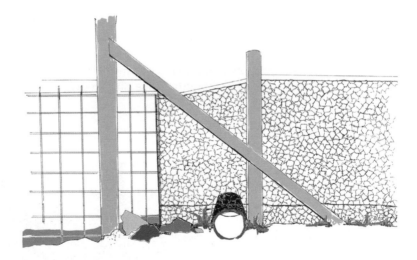

Figure C.21 Beaver-deterrent fencing, including narrow ceramic piping to encourage otter passage but excluding beavers (apart from young kits). Note that such features should be located next to the watercourse and be well secured to ensure that beavers cannot dig around or push through these passages. (Rachael Campbell-Palmer)

Tips
- The lower end of the fence needs to be stiff enough to prevent beavers from forcing it up, and themselves through.

- Young beavers may be able to pass through larger mesh; however, they do not normally attempt to disperse from the natal colony.
- This method relies on the normal behaviour of beavers, and as a result will never be 100% effective, this method should reduce the intervals between colonisation events upstream of the installation, and so the requirement for other measures.
- The design can be adapted to integrate with existing fencing: for example, fencing that excludes rabbits will also exclude beavers, or fencing can also include features that allow otter passage but exclude beavers (Figure C.21).

C.10   Trapping and translocation

It is essential that any requirement to trap beavers is discussed with the appropriate authority and undertaken with highly experienced personnel.

C.10.1   Bavarian beaver traps

Beaver traps should be positioned on foraging trails in areas where fresh field signs are present. Trap use can also be encouraged through appropriate pre-baiting. Care should be taken to ensure that, when in position, the traps are both level and stable, and cannot fall into the water. Traps which are not set on a level surface may not lock properly. Trappers must be aware of any likely fluctuations in the adjacent water levels which would endanger any captured beavers, and must ensure that traps are set away from public interference. Doors and locking mechanisms should be checked and lubricated regularly, especially because asynchrony in the dropping of the doors due to poor lubrication could cause injury to the beaver as it attempts to escape. Commercial vegetable oil is good for this purpose, as it offers less of a scent-deterrent than artificial lubricants. It is advised that latex gloves are worn when handling and setting traps, or the traps should be left, unset, in the elements for a few days to diminish human odours.

Once in position, traps should be set and doors dropped several times to ensure they lock before a trap is set and left. The mesh floor of the trap and treadle should be gently covered with detritus (leaf litter, substrate or bark) from a source nearby to try

Figure C.22 Set Bavarian trap. Note covered trap floor and bait set in middle of treadle.
(R. Campbell-Palmer)

to cover mesh flooring, as beavers tend to be reluctant to step onto uncovered mesh. Care should be taken that none of this substrate impedes the trap doors' descent or interferes with the locking mechanisms. To increase trapping efficiency, a funnel made from nearby branches or wire-mesh panels can be formed to guide the beaver into the trap. Traps should be checked once a day when in operation, ideally being set in the early evening and then checked the following morning. Trap mechanisms and joints can freeze or become stiff, so should be checked regularly and lubricated with vegetable oil as required. A range of trap alarms exist, from simple mobile phones available online to more sophisticated systems (with prices per trap ranging from €10/15 to €100/250 for commercial alarm systems).

Overall, injury rates through this trapping method are very low, and no deaths have been reported in more than 3,000 trappings in Bavaria. However, serious injuries and deaths have been recorded from the use of these traps in captive settings, caused by the door falling down on the beaver's back/neck. Individuals have received nose, claw and teeth injuries while trying to bite and dig their way out of the trap. To reduce this risk, the food, scent or lure must be placed on the trap treadle in the very centre of the trap. None should fall down either side, as beavers may reach in for bait rather than having to enter the trap completely. Animals should not be moved while in the trap, as body parts may become trapped on the floor mesh when setting down the trap.

Potential issues
- Trap placed in an inappropriate location.
- Traps tripping but not latching.
- Traps placed on uneven ground, or at an angle, so that more pressure is required to set off trap.
- Treadle set too high, so that trap is not set off.
- Mesh on trap is too wide, or gauge of wire too light, causing teeth and claw injuries.
- Vegetation, especially on trap floor or side of trap, preventing complete latching.

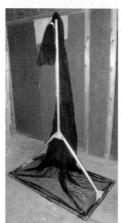

Figure C.23 (Left) Landing net specially designed for boat trapping. Note extending net length to allow beaver to swim upwards into the net; the mouth of the net is then tied behind so that the trapped beaver can be lifted into the boat or onshore. (Right) Quick-release strips or securely tied, non-slip rope can be used to tie nets closed and secure beavers within. (Scottish Beaver Trial)

Figure C.24 Scoop net being used to lift trapped beaver onto shore. This type of net can be used over the side of a boat to scoop up kits. (Scottish Beaver Trial)

Figure C.25 Finer mesh used on landing net to prevent injuries; heavy-duty fishing net, open end tied with non-slip rope. (Scottish Beaver Trial)

Figure C.26 A boat-trapped beaver still within landing net (lying horizontally across boat); the beaver has been covered with hessian to keep it calmer and to keep the torchlight out of its eyes. Note blue quick-release straps have been tied before and after the beaver so it is securely held in the net, so that the animal can be lifted out of boat without biting trappers. (Scottish Beaver Trial)

- Trap-shy animals (those that are very wary or reluctant to enter traps) can be difficult to trap.
- Carnivores, e.g. pine martens, scenting near trap may discourage beavers from entering.
- Interference from people.

C.10.2 Boat trapping

Beavers can be trapped at night from a boat using spotlights and a specially designed landing net to capture adults; long-handled scoop nets can be used for younger animals (<15 kg). Nets with a mesh gauge of ~0.5 cm are recommended to prevent ear tags and claws from becoming snagged. Beavers are spot-lamped from a boat which is then manoeuvred so that the animals can be captured with a net on land or in water of <1 m deep. When appropriate, the trapper at the bow of the boat jumps over the side, directing the landing net over the animal. The landing net must be pushed downwards and contact made with the river bed/ enclosure bottom to ensure that the beaver does not escape. If contact is incomplete, the beaver will often squeeze or dig its way under the net and escape. Care should be taken not to strike the beaver with the net frame in the process. Once secured, the tied end of the net should be let out to allow the beaver to swim to the end, where it often comes to the surface. The open end of the net (mouth end) can then be tied behind the beaver to prevent escape. Care should be taken to avoid possible bite or claw injuries, particularly to the legs and hands. The netted beaver can then be carefully carried to shore or placed in the boat. The beaver's eyes should be covered with a hessian sack or blanket while the animal is being processed and/or crated.

An important modification of these beaver nets is that the end of the net is open. During capture, this is held closed, tied with non-slip rope or fastened with a quick-release buckle tie-down (Figure C.23) which can easily be released, enabling the net to be opened to allow the beaver to be transferred to travel crates or hessian sacks for processing (Rosell and Hovde 2001).

Figure C.27 (Left) Bavarian trap with isolation slide and transport crate lined up, removing the slide then allows the beaver to move itself into the darkened transport crate. (Right) Beavers can also be removed from traps using secured open-ended scooping nets. (R. Campbell-Palmer and Scottish Beaver Trial)

Potential issues

- Damage to boat engine from submerged objects (e.g. rocks, branches).
- Attempted use of method in inappropriate water bodies/use limited to appropriate water bodies.
- Injury of beaver through hitting with net rim.
- Inexperienced trappers may exhibit low trapping success.
- Requires greater consideration of health and safety.

Animal movement

Bavarian traps have option of an internal isolation slide which can confine a captured beaver to one half of the trap. This allows the door at the other end to be opened safely and a capture crate to be inserted, ensuring the easy and safe transport of a captured beaver. Beavers can also be removed from traps using secured open-ended scooping nets. Beaver and net can be placed into the transport crate, and the quick-release or rope tie then released and the net removed from the crate.

C.11 Humane dispatch

Trapped individuals may be released from traps or transported to 'killing pens' (sheet-metal or solid wooden open square pens with sides of ~1 m and 80 cm high). For humane dispatch at close range in a killing pen (or similar), a .22 rim-fire rifle is recommended, for use by experienced operators. It should be noted that rifles are firearms in the UK and as such their use is strictly controlled. Any person using a firearm should ensure that their licence allows them to dispatch beavers.

It is recommended that the shot is administered directly from above, close to perpendicular to a horizontally standing beaver, aiming for between the shoulder blades, so that the shot is likely to disrupt the heart and associated vital structures in proximity (spine, major vessels and airways) at point-blank range without actually making contact

Figure C.28 Simulated beaver cull with .22 rifle in specifically designed 'killing pens' to show target position point between shoulder blades (Bavaria). (R. Campbell-Palmer)

Figure C.29 Norwegian beaver-hunter. (S. Jones)

with the skin. Head shots with .22 rim-fire rifles should be avoided, as bone here is thick and the brain cavity small, with a risk of small-calibre non-jacketed ammunition such as .22 rim-fire deflecting off the cranium without penetrating the skull and/or totally missing the brain. A .410 shotgun may be used at point-blank range, without actually making contact with the surface of the skin, with the aiming point as described above or alternatively firing from behind, aiming from the ear towards the opposite eye. Shot size should be a minimum of BB/No. 1 shot (4.06 mm).

Where dam creation is an issue with a particular family, then the removal of an actively maintained dam structure during the day will often result in a sudden water-level drop which the beavers are likely to attempt to repair at night. They can then be shot by experienced individuals, preferably by centre-fire rifle with a minimum bullet weight of 50 grains (3.24 g), minimum muzzle velocity of 2,450 ft/s (746.76 m/s), minimum muzzle energy of 100 ft/lb (1,356 joules) (SNH 2014) or with a 16-, 20- or 12-bore shotgun with AAA shot (5.16 mm pellets) or larger at a range of 20 m or less. Distances beyond this are not recommended, as, given their thick skin and dense fur, beavers may receive significant injuries which may not be immediately fatal, and they may try to enter water, where they may drown or die later from wounds and/or infection. Shot smaller in diameter than BB/No. 1 shot is not a reasonable or humane method of killing beavers unless used at point-blank range for humane dispatch as described above.

Euthanasia can also be achieved through humane injection by a qualified veterinary surgeon, using sodium pentobarbitone as with other mammals (80–160 mg/kg intravenously via the ventral tail, cephalic or saphenous vein).

Appendix D
Management protocol chart

Beaver activity	Management options		Advantages	Disadvantages	Costs	Solution	Effectiveness
Damming	Flow device	Dam piping	Reduces/sets water level, limits flooding	Installation and occasional maintenance costs, variable success if not installed correctly	Moderate	Long term	Good
		Culvert protection	In-pipe blockage prevention, limits flooding	Installation and ongoing maintenance costs to remove debris as may be regularly blocked	Low	Medium term	Fair
	Dam removal		Reduces water level, limits flooding, no installation	Ongoing monitoring as dam may be rebuilt, increase in tree-felling to rebuild	Low/moderate	Short term	Good
	Dam-notching		Reduces water level, limits flooding, no installation	Ongoing monitoring as dam may be rebuilt, increase in tree-felling to rebuild	Low	Short term	Fair
Burrowing	Burrow management		Prevention of burrow use/extension	Very short-term impact, beavers likely to burrow in the same area	Low	Short term	Fair
	Bank protection		Long-term protection of bank from burrowing from various species	Installation costs, pushes beavers to next unprotected area	High	Long term	Good
Tree browsing	Tree protection	Mesh	Protection and retention of key trees	Installation, occasional maintenance as tree grows, may not be aesthetically suitable	Low/moderate	Long term	Good
		Paint	Protection and retention of key trees	Labour-intensive initially, occasional reapplication, may not be aesthetically suitable	Low/moderate	Medium term	Fair

Management protocol chart – *continued*

Beaver activity	Management options	Advantages	Disadvantages	Costs	Solution	Effectiveness
Activity in undesired area	Exclusion fencing	Long-term protection of large areas of trees	Installation and maintenance costs, variable success if wrong fencing used	Moderate/ high	Long term	Good
	Deterrent fencing	Dissuasion of beavers from dispersing into an upstream area	Installation and maintenance costs, variable success if not installed correctly	High	Long term	Fair
	Electric fencing	Immediate exclusion from an area/site	Effect likely to be temporary, risk of injury/death of beaver	Low/ moderate	Short term	Fair
	Beaver trapping	Temporarily removes beaver activity	Recolonisation of area by beavers likely	Moderate/ high	Short term	Fair

Appendix E
Captive beaver fencing recommendations

Beaver Enclosure Fence Specifications:

The following fence specification is the minimum standard that should be required for the keeping of European beavers in fenced enclosures licensed for the purposes of possession, under the Habitats Regulations, or for conditional release into the wild (within a fenced enclosure), under the Wildlife & Countryside Act.

Fencing

Materials: galvanized high tensile mesh, or similar – with smaller dimension of mesh size no more than 10 cm (4 inch) – e.g. 8cm vertical spacing deer or badger fencing. At lower part of fence the larger dimension of the mesh should also, ideally, not exceed 10cm, though up to 15cm (6 inches) is acceptable (larger mesh size towards the top of the fence – as is often the design of commercially available netting – is acceptable, but this should not exceed 19cm in the larger dimension and still must not exceed 10cm in the smaller dimension). This should preferably be of a locked-joint or weld-mesh jointed type – hinge-joint netting should be avoided as the verticals can be easily pushed out of alignment. Similarly, mild-steel netting is not desirable as it is easily distorted and the mesh bent out of shape.

If the mesh size exceeds 10cm in either dimension in the lower part of the fence it must be 'lined' with an additional layer of wire netting, such as heavy gauge hexagonal mesh, to reduce the effective mesh size to less than 10cm. Hexagonal mesh on its own, however, is not acceptable.

Fence height: minimum 120 cm (4 feet) above ground.

Fence lap: minimum 45 cm (18 inch) skirt of netting securely fixed to the ground on the beaver's side of the fence. This should be of same/similar specification to the lower part of the vertical fence. Ideally, there should also be a section of fencing buried vertically underground, but there should still be a lapped section as described.

Electrified 'outrigger': at least one single-strand electric fence wire should be fitted on the inside (beaver side) of the fence. If this is a single wire, it should be ~15 cm from the ground. Two wires are preferable, with the second 30-40cm above the ground. These must not be allowed to 'short' on the fence netting. Electric fencing must be checked regularly.

Culvert grills

Culverts and other inflow/outflow points are particularly vulnerable to breaches. These should be secured with a metal grill or grid with maximum spacing of 10 cm (4 inch) spacing. The bottom of the metal grill/grid should be fixed in a concrete bed, or similar. Regular clearance of the grill/grid will be a necessary part of the maintenance programme.

C J Wilson
RSAT
February 2009.

Appendix F
Terminology for various beaver-management devices

Developed by Skip Lisle, Beaver Deceivers International, USA.

Beaver Deceivers™ (BD; 1995) are 'flow devices' completely or partly comprising a high-quality, wood-frame fence built on the upstream end of road culverts and other human-made outlets. The frame uses specific materials and has three basic components: posts, a horizontal 'stringer' between posts, and, if necessary, diagonal braces on the posts. Special types of fencing and fasteners are also used. BDs have steadily evolved. Today, they often have relatively small, square three sided fences in front of the culvert combined with an upstream pipe system called a Castor Master™. The initial fence is called a Receiver fence because it 'receives' the pipe system (sometimes, a simple 'straight fence' can replace the three-sided version). Because the topography of every outlet is different, a key concept of BD design is a flexible, mould-to-fit approach. The 'BD' name is popular and often inaccurately used to describe flow devices of *any* kind.

Double-Filter System™ (DFS; 1998) is a general term used to describe flow devices that have some form of fence in front of the culvert (filter #1). A pipe extends upstream from that and is protected, at its intake, by another fence (filter #2). Most Beaver Deceivers™ are double-filter systems.

Trapezoidal Fences™ (1995) are a concept associated with BDs from the late 1990s that were usually large, stand-alone fences (no pipe systems). The idea behind a trapezoid is that, among simple shapes with the same front-to-back length, its sides create the most unnatural direction for a dam (furthest from perpendicular to stream flow). Its perimeter also has the longest length, creating the largest area to 'filter' water away from beavers. They were large because, attached directly to the dam (road), and without 'separation', they were vulnerable. Once good pipe systems began being added in 1998, greatly improving security, outlet fences were less crucial and thus began to become smaller and easier to build. A solid pipe combined with a good filter represents a powerful concept, and is where most of the 'deceiving' takes place. Extending upstream, and devoid of damming stimuli, it essentially moves a dam leak to an incongruous location away from the dam.

Castor Masters™ (CM; 1998) are pipe systems used, in conjunction with fences, in BDs™. They are also used alone in 'regular' beaver dams. They incorporate a variety of different types of pipes, and use Round Fence™ and Square Fence™ filters.

Round Fences™ (RF; 1998) are upright, self-supporting cylinders of wire mesh. They are used either enclosed (top and bottom), or not. They were our first filter – beavers out, water in – designed to protect the ends of pipes in Castor Masters™.

Square Fences™ (SF; 2007) have largely supplanted RFs in recent years. Square or rectangular, they are, in the most basic telling, fencing attached to a wooden stringer at the top. Although not always necessary, posts are usually driven into the substrate in each corner and screwed to the stringer. They have several advantages over RFs.

Swept-Wing Fences™ (SWF; 2003) are designed to turn back beavers that are moving upstream along a narrow, vulnerable watercourse such as an irrigation ditch. 'Area exclusion' is the best approach in a long ditch that beavers could dam quickly at dozens of different points; it would be too expensive to build a new flow device for every potential dam. Where SWF's cross the stream, they have to be combined with a pipe system. Otherwise, if a beaver gets 'inside' that will be the first place to be dammed.

Flow-Device Fish Ladders™ (2002) are small structures placed on the downstream ends of pipes in dams or BDs™ to help let fish through. They are also used on road culverts that block fish passage because they are 'hanging', or perched, on the downstream end.

Turtle-Beaver Doors™ (1998) are components of culvert fences that allow larger animals to traverse streams, and pass through the fence, so they are not forced to go across the road. They also have to be designed to prevent beavers from pulling debris through them. They come in a variety of designs and styles.

Whirlpool Breaks™ (2011) prevent whirlpools from forming on the ends of pipes. If whirlpools, along with their sucking sound, reach the surface, beavers will bury the filter.

Misery Multipliers™ (MM; 2010) are a way of adding security to 6 in. (15 cm) mesh fencing. Small beavers will sometimes pull a lot of debris through these holes and then into the culvert. MMs can be designed in a number of ways, and basically make the economics less favourable by forcing beavers to go through two walls.

BDI Live Traps™ (2012) are built with BD components (wood and steel mesh). They have a simple gravity-door-drop that a beaver triggers once inside. The beaver is not subjected to a violent, spring-loaded door, and has more room to move around when caught. The structure doubles as a carrying case for moving the animal elsewhere.

BDI Tree Guards™ (2006). Our method of protecting trees from chewing by beavers.

BDI Trails, Boardwalks and Decks™ (2009). Unique frame structures and designs for accessing and enjoying wetlands and other beautiful natural areas. Beaver Deceivers International also has expertise in determining the best locations for these products.

Appendix G
Beaver fieldwork risk assessment

Training/supervision	Review site risk assessment and brief personnel on risks and controls before work starts
PPE required	Suitable footwear and clothing to suit conditions, mobile phone

COMMON HAZARDS associated with the work activity	WHO IS AT RISK? Public, staff, trainees, volunteers, children	RISK RATING Risk level before controls are in place	CONTROL MEASURES Existing measures in place before work is allowed to start	RESIDUAL RISK Remaining risk after controls are in place
Working near water/ drowning	Staff, contractors, trainees and volunteers	Low	• Adequate supervision/brief personnel • Buddy system in place • Avoid walking directly along water's edge, especially in areas of deep water • Competent swimmers • PPE as required	Low
Manual handling	Staff, contractors, trainees and volunteers	Medium	• Training/awareness/use of safe lifting techniques • Break down awkward/heavy loads if possible • Rotate tasks to avoid repetitive strain • Seek assistance where required	Low
Slips, trips and falls	Staff, contractors, trainees and volunteers	Medium	• Suitable footwear for weather and site conditions • Plan safe access to and egress from routes • Use lights as appropriate	Low
Wildlife encounters	Staff, contractors, trainees and volunteers	Medium	• Do not corner or closely approach wildlife • Do not attempt to handle animals, especially injured individuals, without licence, experience and appropriate equipment	Low

Hazard	Who at risk		Controls	
Illness and disease – Weil's disease/Lyme disease/tetanus	Staff, contractors, trainees and volunteers	Low	• Keep arms and legs covered, and protect existing injuries • Follow advice for safe removal of ticks • Wash hands before eating, drinking or smoking • Up-to-date tetanus vaccination	Low
Violent encounters – landowners/livestock/ stalking season	Staff, contractors, trainees and volunteers	Low	• Be aware of stalking season/boundaries/livestock • Do not enter any areas without landowner's permission; and follow any specific instructions	Low
Adverse weather	Staff, contractors, trainees and volunteers	Medium	• Suitable clothing for variable conditions • Stop work in extreme/hazardous weather	Low

List any additional hazards and extra controls where those listed above are not sufficient to manage the risk

Result	Likely	Probable	Remote	Improbable
Fatal/disability	High	High	Medium	Low
Major injury	High	High	Medium	Low
Minor injury	Medium	Medium	Low	Low
No injury	Low	Low	Low	Low

Likely – happens repeatedly, expected; Probable – will happen more than once; Remote – unlikely, though conceivable; Improbable – highly unlikely.

NAME	SIGNATURE	DATE

Appendix H
Beaver survey data sheets from the Scottish Beaver Trial

Type:		Feature:		Usage		To include:	Location:
Dw Dwelling		**B** Burrow		**U** Used			**≤5** Along water's edge (up to 5m)
		L Lodge		**D** Disused			**5-10** 5-10m from water's edge
Cn Construction		**Dm** Dam					**10-20** 10-20m from water's edge
		Ca Canal					**≥20** 20m+ from water's edge
FS Feed Sign		**FC** Food cache				Underwater stores of cut saplings and branches outside the lodge/burrow	**OW** Overhanging water
		TBC Tree/branch cutting				Felled tree, Cut branch, stripped branch/stick, Gnawed tree	**IW** In water
		FSt Feeding station					**O** Other
		FT Foraging Trail					
		O Other				Grazed area, Aquatic macrophyte mats	
A Activity		**T** Tracks					
		SM Scent mound or marking				Note whether single mark or recent marking of frequently used mound	

SBT Field Signs Survey

Survey date:

Locations Surveyed:

GPS Waypoint	Eastings	Northings	Type	Feature	Usage	Age	Location	Comment	Image ID	Recorders

BEHAVIOURAL OBSERVATION & LOCATION FIELD SHEET **ALL Fields must be completed**

Date:

Observer(s):

Location:

Start time: | Finish time:

Behaviour code	Fix #***	Time (HH:MM)	Animal	Interacting animal	Light Y/N	Disturb²	Comments

BEHAVIOUR CODES MUST BE AS FOLLOWS (see Appendix D ethogram for descriptions of behaviours):

Locomotion:	**Sw** Swim, **Wa** Walk, **D** Dive
Forage:	**F** Forage. **E** Eat. AND Forage type: **a** aquatic, **h** herb, **w** wood e.g. **Eh** = Eat herb/Fw = Forage woody
Territorial:	**Sm** Scent-mark, **Fi** Fight, **SD** Stick-display
Social:	**NN** Nose-nose, **Wr** Wrestle, **AG** Allogroom, **C** Caravan
Misc:	**B** Build, **G** Groom, **S** Sit/Still, **A** Alert, **P** Provision, **T** Tail-slap, **O** Other

²**Disturbance score – write in number**

1. Animal obviously disturbed, changed location and/or changed behaviour

2. Animal possibly disturbed, behaviour may have been affected

3. Animal not disturbed, and behaving normally

Lodge/Den counts Field Sheet **ALL Fields must be completed**

Month/year:

Loch	Lodge/Den		Date	Observer(s)	No. adults	No. sub adults	No. yearlings	No. kits	Comments
	Easting	Northing							

1 sheet to be completed per month.

Beaver Health Status *To be completed by trained staff only*

Date	Beaver	Body Condition	Behaviour	Health Status	Comments	Recorder

Body Condition/Health Status: Poor, ok good, unsure, not seen.   **Behaviour:** normal, abnormal, not sure.

References

Åhlén, P.A. (2001) The parasitic and commensal fauna of the European beaver (*Castor fiber*) in Sweden. Honours thesis, Department of Animal Ecology, Swedish University of Agricultural Sciences.

Alexander, M.D. (1998) Effects of beaver (*Castor canadensis*) impoundments on stream temperature and fish community species composition and growth in selected tributaries of Miramichi River, New Brunswick. Department of Fisheries and Oceans, Science Branch, Maritimes Region, Gulf Fisheries Centre.

Allen, A.W. (1983) Habitat suitability index models: beaver. US Fish and Wildlife Service FWS/0BS-82/10.30.

Avery, E.L. (1983) A bibliography of beaver, trout, wildlife, and forest relationships with special references to beaver and trout. Technician Bulletin 137. Wisconsin Department of Natural Resources. Madison, WI.

Baker, B.W., Peinetti, H.R., Coughenour, M.C. and Johnson, T.L. (2012) Competition favors elk over beaver in a riparian willow ecosystem. *Ecosphere* 3 (art 95): 1–15.

Balodis, M.M. (1994) Beaver population of Latvia: history, development and management. *Proceedings of the Latvian Academy of Sciences B* no. 7/8 (564/565): 122–127.

Barlow, A.M., Gottstein, B. and Mueller, N. (2011) *Echinococcus multilocularis* in an imported captive European beaver (*Castor fiber*) in Great Britain. *Veterinary Record* doi: 10.1136/vr.d4673.

Barnes, W.J. and Dibble, E. (1988) The effects of beaver in riverbank forest succession. *Canadian Journal of Botany*, 66: 40 44.

Baskin, L. and Sjöberg, G. (2003) Planning, coordination and realisation of Northern European beaver management, based on the experience of 50 years of beaver restoration in Russia, Finland and Scandinavia. *Lutra* 46: 243–250.

Batbold, J., Batsaikhan, N., Shar, S., Amori, G., Hutter, R., Kryštufek, B., Yigit, N., Mitsain, G. and Muñoz, L.J.P. (2008) *Castor fiber*. In *IUCN Red list of Threatened Species*. www.iucnredlist.org/details/4007

Batty, D. (2002) Beavers: aspen heaven or hell? In Cosgrove, P. and Amphlett, A. (eds) *The Biodiversity and Management of Aspen Woodlands. Proceedings of a One-day Conference Held in Kingussie, Scotland, 25 May 2001*, The Cairngorms Biodiversity Action Plan, Grantown-on-Spey, UK, pp. 41–44.

Beck, A. and Hohler, P. (2000) Einsatz von künstlichen Biberbauten. *Ingenieurbiologie* 1/00: 26–29.

Beedle, D.L. (1991) Physical dimensions and hydrologic effects of beaver ponds on Kuiu Island in southeast Alaska. MSc thesis, Oregon State University, Corvallis.

Belfiore, N.M. (2006) Observation of a beaver beetle (*Platypsyllus castoris* Ritsema) on a North American river otter (*Lontra canadensis* Schreber) (Carnivora: Mustelidae: Lutrinae) in Sacramento county, California (Coleoptera: Leiodidae: Platypsyllinae). *The Coleopterists Bulletin* 60: 312–313.

Bhat, M.G., Huffaker, R.G. and Lenhart, S.M. (1993) Controlling forest damage by dispersive beaver populations: centralized optimal management strategy. *Ecological Applications* 3: 518–530.

Bobba, A.G., Dean, S. and Singh, V.P. (1999) Sensitivity of hydrological variables in the northeast Pond River watershed, Newfoundland, Canada, due to atmospheric change. *Water Resources Management*, 13: 171–188.

Bochkov, A.V. and Saveljev, A.P. (2012) Fur mites of the genus *Schizocarpus Trouessart* (*Acari: Chirodiscidae*) from the Eurasian beaver *Castor fiber tuvinicus Lavrov* (*Rodentia: Castoridae*) in the Azas River (Tuva Republic, Russia). *Zootaxa* 3410: 1–18.

Boyles, S.L. (2006) Report on the efficacy and comparative costs of using flow devices to resolve conflicts with North American beavers along roadways in the coastal plain of Virginia. Christopher Newport University, VA, USA.

Boyles, S.L. and Savitzky, B.A. (2009) An analysis of the efficacy and comparative costs of using flow devices to resolve conflicts with North American beavers along roadways in the coastal plain of Virginia. Road Ecology Center, University of California, Davis.

Bräuer, I. (2002) *Artenschutz aus volkswirtschaftlicher Sicht: eine Nutzen-Kostenanalyse der Biberwiedereinbürgerung in Hessen*. Marburg: Metropolis Verlag.

Brooks, R.P., Fleming, M.W. and Kenelly, J.J. (1980) Beaver colony response to fertility control: evaluating a concept. *Journal of Wildlife Management* 44: 568–575.

Bryant, M.D. (1984) The role of beaver dams as coho salmon habitat in southeastern Alaska streams. In: Walton, J.M. and Houston, D.D. (eds), *Proceedings of the Olympic Wild Fish Conference*. Peninsula College, Fisheries Technology Program, Port Angeles, WA, pp. 183–192.

BSWG (2015) Final report of the Beaver Salmonid Working Group. Prepared for the National Species Reintroduction Forum, Inverness.

Buech, R.R. (1983) Modification of the Bailey live trap for beaver. *Wildlife Society Bulletin*, 1: 66–68.

Bulte, E.H. and Rondeau, D. (2005) Why compensating wildlife damages may be bad for conservation. *Journal of Wildlife Management* 69: 14–19.

Bulter, D.R. and Malanson, G.P. (2005) The geomorphic influences of beaver dams and failures of beaver dams. Geomorphology 71: 48-60.

Bush, A.O. and Samuel, W.M. (1981) A review of helminth communities in beaver (*Castor* spp.) with a survey of *Castor canadensis* in Alberta, Canada. In Chapman, J.A. and Pursley, D. (eds) *Proceedings of Worldwide Furbearer Conference*, 3–11 August, Frostburg, pp. 678–689.

Busher, P.E., Warner, R.J. and Jenkins, S.H. (1983) Population density, colony composition, and local movements in two Sierra Nevadan beaver populations. *Journal of Mammalogy*, 64: 314–318.

Butler, D.R. and Malanson, G.P. (2005) The geomorphic influences of beaver dams and failures of beaver dams. *Geomorphology* 71: 48–60.

Callahan, M. (2003) Beaver management study. *Association of Massachusetts Wetland Scientists (AMWS) Newsletter* 44: 12–15.

Callahan, M. (2005) Best management practices for beaver problems. *Association of Massachusetts Wetland Scientists (AMWS) Newsletter* 53, 12–14.

Campbell, R. and Tattersall, F. (2003) The Ham Fen Beaver Project. First report. Wildlife Conservation Research Unit, University of Oxford.

Campbell, R.D. (2006) What has the beaver got to do with the freshwater mussel decline? A response to Rudzīte (2005). *Acta Universitatis Latviensis*, 710: 139–140.

Campbell, R.D. (2010) Demography and life history of the Eurasian Beaver *Castor fiber*. DPhil thesis, University of Oxford.

Campbell, R.D., Rosell, F., Nolet, B.A. and Dijkstra V.A.A. (2005) Territory and group size in Eurasian beavers (*Castor fiber*): echoes of settlement and reproduction. *Behaviour Ecology and Sociobiology* 58: 597–607.

Campbell, R.D., Dutton, A. and Hughes, J. (2007) Economic impacts of the beaver. Report for the Wild Britain Initiative. Wildlife Conservation Research Unit, University of Oxford. http://www.scottishbeavers.org.uk/docs/003__021__general__Campbell_et_al_2007_Economic_impacts_of_the_beaver__1282729674.pdf.

Campbell, R.D., Harrington, A., Ross, A. and Harrington, L. (2012a) Distribution, population assessment and activities of beavers in Tayside. Scottish Natural Heritage Commissioned Report No. 540.

Campbell, R.D., Nouvellet, P., Newman, C., Macdonald, D.W. and Rosell, F. (2012b) The influence of mean climate trends and climate variance on beaver survival dynamics. *Global Change Biology* 18: 2730–2742.

Campbell, R.D., Newman, C., Macdonald, D.W. and Rosell, F. (2013) Proximate weather patterns and spring green-up phenology effect Eurasian beaver (*Castor fiber*) body mass and reproductive success: the implications of climate change. *Global Change Biology* 19: 1311–1324.

Campbell, R.D., Rosell. F., Newman, C. and Macdonald, D.W. In press. Resource history influences age-related somatic condition and onset of reproductive senescence in the Eurasian Beaver. *PLOS ONE*.

Campbell-Palmer, R. and Rosell, F. (2010) Conservation of the Eurasian beaver *Castor fiber*: an olfactory perspective. *Mammal Review* 40: 293–312.

Campbell-Palmer, R. and Rosell, F. (eds) (2013) *Captive Management Guidelines for Eurasian Beaver* (Castor fiber). RZSS, BookPrintingUK.com.

Campbell-Palmer, R., Girling, S., Rosell, F., Paulsen, P. and Goodman, G. (2012) Echinococcus risk from imported beavers. *Veterinary Record* 3 March, 235.

Campbell-Palmer, R., Girling, S., Pizzi, R., Hamnes, I.S., Øines, Ø. and del Pozo, J. (2013) *Stichorchis subtriquetrus* in a free-living beaver in Scotland. *Veterinary Record* doi: 10.1136/vr.101591.

Campbell-Palmer, R., Pizzi, R., Dickinson, H. and Girling, S. (2015) Trapping and health screening of free-living beaver within the River Tay catchment, east Scotland. Scottish Natural Heritage Commissioned Report, No. 681, Battleby.

Carr, W.H. (1940) Beavers and birds. *Bird Lore* 42: 141–146.

Chanin, P. (2006) Otter road casualties. *Hystrix, Italian Journal of Mammalogy* 17: 79–90.

Cirmo, C.P. and Driscoll, C.T. (1993) Beaver pond biogeochemistry: acid neutralizing capacity generation in a headwater wetland. *Wetlands* 13: 277–292.

Close, T.L. (2003) Modifications to the Clemson Pond Leveler to facilitate Brook trout passage. Minnesota Department of Natural Resources Special Publication 158.

Coles, B.J. (2006) *Beaver in Britain's Past*. WARP Occasional Papers (Book 19). Oxford: Oxbow Books.

Collen, P. (1997) Review of the potential impacts of re-introducing Eurasian beaver *Castor fiber* L. on the ecology and movement of native fishes, and the likely implications for current angling practices in Scotland. Scottish Natural Heritage Commission Report No. 86.

Collen, P. and Gibson, R.J. (2000) The general ecology of beavers (*Castor spp.*), as related to their influence on stream ecosystems and riparian habitats, and the subsequent effects on fish – a review. *Reviews in Fish Biology and Fisheries* 10: 439–461.

Collett, R. (1897) Bœveren I Norge, dens Utbredelsen og Levemaade (1896). *Bergens Museums Aarbog* 1: 1-139.

Conroy, J.W.H. and Kitchener, A.C. (1996) The Eurasian beaver (*Castor fiber*) in Scotland: a review of the literature and historical evidence. Scottish Natural Heritage Commission Report No. 49.

Cook, D.B. (1940) Beaver–trout relations. *Journal of Mammalogy* 21(4): 397–401.

Cope, D., Pettifor, R., Griffin, L. and Rowcliffe, J. (2003) Integrating farming and wildlife conservation: the Barnacle Goose Management Scheme. *Biological Conservation* 110: 113–122.

Cosgrove, P.J., Young, M.R., Hastie, L.C., Gaywood, M. and Boon, P.J. (2000) The status of the freshwater pearl mussel *Margaritifera margaritifera* Linn. in Scotland. *Aquatic Conservation: Marine and Freshwater Ecosystems* 10: 197–208.

Cross, H.B., Campbell-Palmer, R., Girling, S. and Rosell, F. (2012) The Eurasian beaver (*Castor fiber*) is apparently not a host to blood parasites in Norway. *Veterinary Parasitology* 190: 246–248.

Cunjak, R.A. (1996) Winter habitat of selected stream fishes and potential impacts from land-use activity. *Canadian Journal of Fisheries and Aquatic Science* 53: 267–282.

Cunjak, R.A. and Therrien, J. (1998) Inter-stage survival of wild juvenile Atlantic salmon, *Salmo salar* L. *Fisheries Management and Ecology* 5: 209–223.

Czech, A. and Lisle, S. (2003) Understanding and solving the beaver (*Castor fiber* L.)–human conflict: an opportunity to improve the environment and economy of Poland. *Denisia* 9: 91–98.

Danilov, P.I. and Kan'shiev, V.Y. (1983) The state of populations and ecological characteristics of European (*Castor fiber* L.) and Canadian (*Castor canadensis* Kuhl.) beavers in the north-western USSR. *Acta Zoological Fennica* 174: 95–97.

Danilov, P.I., Kan'shiev, V.Y. and Fyodorov, F. (2011) Characteristics of North American and Eurasian beaver ecology in Karelia. In Sjöberg, G. and Ball, J.P. (eds) *Restoring the European Beaver: 50 Years of Experience*. Sofia: Pensoft Publishers, pp. 55–72.

Davic, R.D. (2003) Linking keystone species and functional groups: a new operational definition of the keystone species concept. *Conservation Ecology* 7(1): r11.

Davidson, R., Øines, Ø. and Nortstrom, M. (2009) Surveillance and control programme for *Echinococcus multilocularis* in red foxes (*Vulpes vulpes*) in Norway. Annual report 2009. In Kardsson, A.C., Jordsymr, H.M., Hellberg, H. and Svilbard, S. (eds), *Surveillance and Control Programmes for Terrestrial and Aquatic Animals in Norway*. National Veterinary Institute.

Davis, J.R. (1984) Movement and behaviour patterns of beaver in the Piedmont of South Carolina. Thesis, Clemson University, SC.

Deblinger, R.D., Woytek, W.A. and Zwick, R.R. (1999) Demographics of voting on the 1996 Massachusetts ballot referendum. *Human Dimensions of Wildlife* 4: 40–55.

DEFRA (2005) Making space for water: taking forward a new government strategy for flood and coastal erosion risk management in England. First response.

DEFRA (2008) Future Water: The Government's Water Strategy for England.

DEFRA (2012) What is the risk of introducing *Echinococcus multilocularis* to the UK wildlife population by importing European beavers which subsequently escape or are released? http://www.defra.gov.uk/animal-diseases/files/qra-non-native-species-echinoccocus-120627.pdf (accessed 4 July 2012).

Dewas, M., Herr, J., Schley, L., Angst, C., Manet, B., Landry, P. and Catusse, M. (2012) Recovery and status of native and introduced beavers *Castor fiber* and *Castor canadensis* in France and neighbouring countries. *Mammal Review* 42: 144–165.

Dickie, I., Hughes, J. and Esteban, A. (2006) 'Watched Like Never Before …': The Local Economic Benefits of Spectacular Bird Species. Sandy: RSPB.

Donkor, N.T. and Fryxell, J.M. (1999) Impact of beaver foraging on structure of lowland boreal forests of Algonquin Provincial Park, Ontario. *Forest Ecology and Management* 118: 83–92.

Duncan, S.L. (1984) Leaving it to beaver. *Environment* 26: 41–45.

Duff, A.G., Campbell-Palmer, R. and Needham, R. (2013) The beaver beetle *Platypsyllus castoris* Ritsema (Leiodidae: Platypsyllinae) apparently established on reintroduced beavers in Scotland, new to Britain. *The Coleopterist* 22: 9–19.

DVWK (Deutscher Verband für Wasserwirtschaft und Kulturbau e.V.) (eds) (1997) *Gestaltung und Sicherung der von Bisam, Biber und Nutria besiedelten Ufer, Deiche und Dämme*. Bonn: Wirtschafts und Verlagsgesellschaft Gas und Wasser GmbH, pp. 1–83.

Dzięciołowski, R.M. and Gozdziewski, J. (1999) Beaver management in the Baltic states. In Busher, P.E. and Dzięciołowski, R.M. (eds) *Beaver Protection, Management, and Utilization in Europe and North America*. New York: Kluwer Academic/Plenum, pp. 25–30.

Eckert, J. (1997) Epidemiology of *Echinococcus multilocularis* and *E. granulosus* in central Europe. *Parasitologia* 39: 337–344.

Eckert, J. and Deplazes, P. (1999) Alveolar echinococcosis in humans: the current situation in central Europe and need for countermeasures. *Parasitology Today* 15: 315–319.

Eckert, J., Conraths, F.J. and Tackmann, K. (2000) *Echinococcosis*: an emerging or re-emerging zoonosis? *International Journal for Parasitology* 30: 1283–1294.

Elliott, M. and Burgess, P. (2013) *The Devon Beaver Project. The Story so Far …* Exeter: Devon Wildlife Trust. http://www.devonwildlifetrust.org/i/Beaver_report_27-8-13.pdf

Elmeros, M., Madsen, A.B. and Berthelsen, J.P. (2003) Monitoring of reintroduced beavers in Denmark. *Lutra* 46: 153–162.

Environment Agency (2003) *River Habitat Survey in Britain and Ireland: Field Survey Guidance Manual.* www.riverhabitatsurvey.org/wp-content/uploads/2012/07/RHS_1.pdf.

Environment Agency (2008a) Delivery of making space for water: the role of land management in delivering flood risk management. Final report.

Environment Agency (2008b) The Environment Agency's position on land management and flood risk management, position statement. Final version.

EPIC (2015) Public health risk of Giardia and Cryptosporidium posed by reintroduced beavers into Scotland. Centre of Expertise on Animal Disease Outbreaks. http://www.snh.gov.uk/docs/A1832817.pdf

Ermala, A. (1997) On beaver hunting and its influence on the beaver population in Finland. In Nitsche, K.A. and Pachinger, K. (eds) *Proceedings of the 1st European Beaver Symposium,* Bratislava, Slovakia, 15–19 September 1997, Slovak Zoological Society, pp. 23–26.

Ermala, A. (2001) The Finnish beaver status at present and means of controlling it. In Czech, A. and Schwab, G. (eds), *The European Beaver in a New Millennium: Proceedings of the 2nd European Beaver Symposium, 27–30 September 2000, Kraków, Poland,* pp. 161–163.

Forestry Research (2014) *Slowing the Flow at Pickering.* Crown copyright. http://www.forestry.goc.uk/website/forestresearch.nsf/ByUnique/INFD-7YML5R

France, R.L. (1997) The importance of beaver lodges in structuring littoral communities in boreal headwater lakes. *Canadian Journal of Zoology/Revue Canadienne de Zoologie* 75: 1009–1013.

Fryxell, J.M. (2001) Habitat suitability and source-sink dynamics of beavers. *Journal of Animal Ecology* 70: 310–316.

Fur Institute of Canada (2014) Certified Traps – Agreement on International Humane Trapping Standards (AIHTS) Implementation in Canada, updated 6 October 2014. Ottawa: Fur Institute of Canada (*www.fur.ca*).

Fustec, J., Lode, T., Le Jacques, D. and Cormier, J.P. (2001) Colonization, riparian habitat selection and home range size in a reintroduced population of European beavers in the Loire. *Freshwater Biology* 46: 1361–1371.

Gaydos, J.K., Zabek, E. and Raverty, S. (2009) *Yersinia pseudotuberculosis* septicaemia in a beaver from Washington State. *Journal of Wildlife Disease* 45: 1182–1186.

Genney, D.R. (2015) The Scottish Beaver Trial: lichen impact assessment 2010–2014, final report. Scottish Natural Heritage Commissioned Report No. 694.

Giraldus Cambrensis (1180) *The Itinerary of Archbishop Baldwin through Wales.* Project Gutenberg Etexts 1997. http://www.gutenberg.org/cache/epub/1148/pg1148.html

Girling, S.J., Campbell-Palmer, R., Pizzi, R., Fraser, M., Cracknell, J., Arnemo, J. and Rosell, F. (2015) Haematology and serum biochemistry parameters and variations in the Eurasian beaver (*Castor fiber*). *PLOS ONE* 12: 10(6):e0128775. doi: 10.1371/journal.pone.0128775.

Goodman, G., Girling, S., Pizzi, R., Rosell, F. and Campbell-Palmer, R. (2012) Establishment of a health surveillance program for the reintroduction of the Eurasian beaver (*Castor fiber*) into Scotland. *Journal of Wildlife Disease* 48: 971–978.

Gorshkov, D. (2003) Is it possible to use beaver building activity to reduce lake sedimentation? *Lutra* 46: 189–196.

Gorshkov, Y., Easter-Pilcher, A., Pilcher, B. and Gorshkov, D. (1999) Ecological restoration by harnessing the work of beaver. In Busher, P.E. and Dzięciołowski, R.M. (eds), *Beaver Protection, Management and Utilization in Europe and North America.* New York: Kluwer Academic/Plenum, pp. 67–76.

Gorshkov, Y.A., Gorshkov, D.Y., Easter-Pilcher, A.L. and Pilcher, B.K. (2002) First results of beaver (*Castor fiber*) reintroduction in Volga–Kama National Nature Zapovednik (Russia). *Folia Zoologica* 51: 67–74.

Gottstein, B., Frey, C.F., Campbell-Palmer, R., Pizzi, R., Barlow, A., Hentrich, B., Posautz, A. and Ryser-Degiorgis, M.-P. (2014) Immunoblotting for the serodiagnosis of alveolar echonococcosis in live and dead Eurasian beavers (*Castor fiber*). *Veterinary Parasitology* 205: 113–118.

Gow, D. and Elliott, M. (2014) The role of beaver-generated landscapes in flood prevention. *In Practice* 84: 28–33.

Graf, P.M., Wilson, R.P., Qasem, L. Hackländer, K., Rosell, F. (2015) The use of acceleration to code for animal behaviours: a case study in free-ranging Eurasian beavers *Castor fiber*. *PLOS ONE* 10: e0136751.

Grasse, J.E. (1951) Beaver ecology and management in the Rockies. *Journal of Forestry* 49: 3–6.

Grasse, J.E. and Putnam, E.F. (1950) Beaver management and ecology in Wyoming. *Wyoming Game and Fish Commission Bulletin* 6.

Green, K.C. and Westbrook, C.J. (2009) Changes in riparian area structure, channel hydraulics and sediment yield following loss of beaver dams, British Columbia. *Journal of Ecosystems and Management* 10: 68–79.

Grover, A.M. and Baldassarre, G.A. (1995) Bird species richness within beaver ponds in south-central New York. *Wetlands* 15: 108–118.

Guardian (2013) Beaver kills man in Belarus. http://www.theguardian.com/world/2013/may/29/beaver-kills-man-belarus.

Gurnell, A.M. (1998) The hydrogeomorphological effects of beaver dam-building activity progress. *Physical Geography* 22: 167–189.

Gurnell, J., Demeritt, D., Lurz, P.W.W., Shirley, M.D.F., Rushton, S.P., Faulkes, C.G., Nobert, S. and Hare, E.J. (2009) The feasibility and acceptability of reintroducing the European beaver to England. Report prepared for Natural England and the People's Trust for Endangered Species.

Haarberg, O. and Rosell, F. (2006) Selective foraging on woody plant species by the Eurasian beaver (*Castor fiber*) in Telemark, Norway. *Journal of Zoology* 270: 201–208.

Hacking, M.A. and Sileo, L. (1974) *Yersinia enterocolitica* and *Yersina pseudotuberculosis* from wildlife in Ontario. *Journal of Wildlife Disease* 10: 452–457.

Hägglund, Å. and Sjöberg, G. (1999) Effects of beaver dams on the fish fauna of forest streams. *Forest Ecology and Management* 115: 259–266.

Hahn, D. (2003) *The Tower Menagerie: Being the Amazing True Story of the Royal Collection of Wild and Ferocious Beasts.* London: Simon & Schuster.

Halley, D. (1995) The proposed re-introduction of the beaver to Britain. *Reintroduction News* 10: 17–18.

Halley, D. (2011) Sourcing Eurasian beaver *Castor fiber* stock for reintroductions in Great Britain and Western Europe. *Mammal Review* 41: 40–53.

Halley, D.J. and Lamberg, A. (2001) Populations of juvenile salmon and trout in relations to beaver damming of spawning streams. In Czech, A. and Schwab, G. (eds) *The European Beaver in a New Millennium. Proceedings of the 2nd Beaver Symposium, 27–30 September 2000, Kraków, Poland,* pp. 122–127.

Halley, D.J. and Rosell, F. (2002) The beaver's reconquest of Eurasia: status, population development and management of a conservation success. *Mammal Review* 32: 153–178.

Halley, D.J., Jones, A.C.L., Chesworth, S., Hall, C., Gow, D., Jones-Parry, R. and Walsh, J. (2009) The reintroduction of the Eurasian beaver *Castor fiber* to Wales: an ecological feasibility study. NINA Report 457.

Halley, D., Rosell, F. and Saveljev, A. (2012) Population and distribution of Eurasian beaver (*Castor fiber*). *Baltic Forestry* 18: 168–175.

Halley, D., Teurlings, I., Welsh, H. and Taylor, C. (2013) Distribution and patterns of spread of recolonising Eurasian beavers (*Castor fiber* Linnaeus 1758) in fragmented habitat, Agdenes peninsula, Norway. *Fauna Norvegica* 32: 1–12.

Hamilton, A. and Moran, D. (2015) Tayside beaver socio-economic impact study. Scottish Natural Heritage Commissioned Report No. 805.

Härkönen, S. (1999) Forest damage caused by the Canadian beaver (*Castor canadensis*) in South Savo, Finland. *Silva Fennica* 33: 247–259.

Harrington, L.A., Feber, R., Raynor, R. and Macdonald, D.W. (2015) The Scottish Beaver Trial: ecological monitoring of the European beaver *Castor fiber* and other riparian mammals 2009–2014, final report. Scottish Natural Heritage Commissioned Report No. 685.

Hartman, G. (1994) Long-term population development of a reintroduced beaver (*Castor fiber*) population in Sweden. *Conservation Biology* 8: 713–717.

Hartman, G. (1995) Patterns of spread of a reintroduced beaver *Castor fiber* population in Sweden. *Wildlife Biology* 1: 97–103.

Hartman, G. (1999) Beaver management and utilization in Scandinavia. In Busher, P.E. and Dzięciołowski, R.M. (eds) *Beaver Protection, Management, and Utilization in Europe and North America.* New York: Kluwer Academic/Plenum, pp. 1–6.

Hartman, G. (2011) The beaver (*Castor fiber*) in Sweden. In Sjöberg, G. and Ball, J.P. (eds) *Restoring the European Beaver: 50 Years of Experience.* Sofia: Pensoft Publishers, pp. 13–18.

Hartman, G. and Törnlöv, S. (2006) Influence of watercourse depth and width on beaver dam building. *Journal of Zoology* 268: 127–131.

Hay, K.G. (1958) Beaver census methods in the Rocky mountain region. *Journal of Wildlife Management* 22: 395-402.

Herr, J. and Rosell, F. (2004) Use of space and movement patterns in monogamous adult Eurasian beavers (*Castor fiber*). *Journal of Zoology London* 262: 257–264.

Hilfiker, E.L. (1991) *Beavers, Water, Wildlife and History.* New York/Interlaken: Windswept Press.

Hill, E.P. (1982) Beaver (*Castor canadensis*). In Chapman, J.A. and Feldhamer, G.A. (eds) *Wild Mammals of North America: Biology, Management and Economics.* Baltimore/London: John Hopkins University Press.

Hodgdon, H.E. (1978) Social dynamics and behaviour within an unexploited beaver (*Castor canadensis*) population. Dissertation, University of Massachusetts, Amherst.

Hood, G.A. (2012) Biodiversity and ecosystem restoration: beavers bring back balance to an unsteady world. Abstract, 6th International Beaver Symposium, Croatia, 17–20 September.

Hood, G.A. and Bayley, S.E. (2008) Beaver (*Castor canadensis*) mitigate the effects of climate on the area of open water in boreal wetlands in western Canada. *Biological Conservation* 141: 556–567.

Hood, G.A. and Bayley, S.E. (2008) A comparison of riparian plant community responses to herbivory by beaver (*Castor canadensis*) and ungulates in Canada's boreal mixed-wood forest. *Forest Ecology and Management* 258: 1979–1989.

Horn, S., Durka, W., Wolf, R., Ermala, A., Stubbe, A., Stubbe, M. and Hofreiter, M. (2011) Mitochondrial genomes reveal slow rates of molecular evolution and the timing of speciation in beavers (*Castor*), one of the largest rodent species. *PLOS ONE* 6: e14622.

Houston, A.E., Pelton, M.R. and Henry, R. (1995) Beaver immigration into a control area. *Journal of Applied Forestry* 19: 127–130.

Howard, R.J. and Larson, J.S. (1985) A stream habitat classification system for beaver. *Journal of Wildlife Management* 49: 19–25.

Hunter, H.C. (1976) Proposed beaver removal in Cottonwood Creek. Environmental Analysis Report, US Forest Service, Inyo National Forest, White Mountain Ranger District, California, 7 May.

Iason, G.R., Sim, D.A., Brewer, M.J. and Moore, B.D. (2014) The Scottish Beaver Trial: woodland monitoring 2009–2013, final report. Scottish Natural Heritage Commissioned Report No. 788.

IUCN/SSC (2013) *Guidelines for Reintroductions and Other Conservation Translocations*, Version 1.0. Gland, Switzerland: IUCN Species Survival Commission.

Jaczewski, Z. *et al.* (1995). Hand rearing young beavers in captivity. *Proceedings of the International Union of Game Biologists. The Game and the Man XXII.* Sofia, Bulgaria.

Jameson, L.J., Logue, C.H., Atkinson, B., Baker, N., Galbraith, S.E., Carroll, M.W., Brooks, T. and Hewson, R. (2013) The continued emergence of hantaviruses: isolation of a Seoul virus implicated in human disease, United Kingdom, October 2012. *Euro Surveillance* 18: 20344.

Janovsky, M., Bacciarini, L., Sager, H., Gröne, A. and Gottstein, B. (2002) *Echinococcus multilocularis* in a European beaver from Switzerland. *Journal of Wildlife Disease* 38: 618–620.

Janzen, D.H. (1963) Observations on populations of adult beaver beetles, *Platypsyllus castoris* (Platypsyllidae: Coleoptera). *Pan-Pacific Entomologist* 34: 215–228.

Jenkins, S.H. (1975) Food selection by beavers. *Oecologia* 21: 157–173.

Jensen, P.G., Curtis, P.D. and Hameline, D.L. (1999) *Managing Nuisance Beavers along Roadside: A Guide for Highway Departments.* Ithaca, NY: Cornell University.

Johnston, C.A. and Naiman, R.J. (1990) Browse selection by beaver: effects on riparian forest composition. *Canadian Journal of Forestry Research* 20: 1036–1043.

Jones, A., Gilvear, D. and Willby, N. and Gaywood, M. (2009) Willow (*Salix* spp.) and aspen (*Populus tremula*) regrowth after felling by the Eurasian beaver (*Castor fiber*): implications for riparian woodland conservation in Scotland. *Aquatic Conservation: Marine and Freshwater Ecosystems* 19: 75–87.

Jones, C.G., Lawton, J.H. and Shachak, M. (1994) Organisms as ecosystem engineers. *Oikos* 69: 373–386.

Jones, C.G., Lawton, J.H. and Shachak, M. (1997) Positive and negative effects of organisms as physical ecosystem engineers. *Ecology* 78: 1946–1957.

Jones, S. and Campbell-Palmer, R. (2014) The Scottish Beaver Trial: the story of Britain's first licensed release into the wild. Final Report 2014. http://scottishbeavers.org.uk/docs/003_143__scottishbeavertrialfinalreport_dec2014_1417710135.pdf.

Jones, S., Gow, D., Lloyd Jones, A. and Campbell-Palmer, R. (2013) The battle for British beavers. *British Wildlife* 24: 381–392.

Jonker, S.A., Muth, R.M., Organ, J.F., Zwick R.R. and Siemer, W.F. (2006) Experiences with beaver damage and attitudes of Massachusetts residents toward beaver. *Wildlife Society Bulletin* 34: 1009–1021.

Kauhala, K. and Turkia, T. (2013) Majavien elinympäristönkäyttö: alkuperäislajiin ja vieraslajin alustavaa vertailua (summary: habitat use of beavers: preliminary comparison between a native and alien species). *Suomen Riista* 59: 20-33.

Kemp, P.S., Worthington, T.A. and Langford, T.E.L. (2010) A critical review of the effects of beavers upon fish and fish stocks. Scottish Natural Heritage Commissioned Report No. 349.

Kemp, P.S., Worthington, T.A., Langford, T.E.L., Tree, A.R.J. and Gaywood, M.J. (2012) Qualitative and quantitative effects of reintroduced beavers on stream fish. *Fish and Fisheries* 13: 158–181.

Kile, N.B., Nakken, P.J., Rossell, F. and Espeland, S. (1996) Red fox, *Vulpes vulpes* kills beaver, *Castor fiber*, kit. *Canadian Field Naturalist* 110: 338–339.

Kingston, D. (2004) The 2003 Upper Kitwanga Beaver Dam Breaching Program. Gitanyow Fisheries Authority. http://www.skeenafisheries.ca/?page_id=1106.

Kitchener, A.C. and Conroy, J.W.H. (1997) The history of the Eurasian beaver *Castor fiber* in Scotland. *Mammal Review* 27: 95–108.

Knudsen, G.J. (1962) Relationship of beavers to forest, trout and wildlife in Wisconsin. Technical Bulletin 25, Wisconsin Conservation Department, Madison.

Koenen, K., DeStefano, S., Henner, C. and Beroldi, T. (2005) Capturing beavers in box traps. *Wildlife Society Bulletin* 33: 1153–1159.

Kosmider, R., Paterson, A., Voas, A. and Roberts, H. (2013) *Echinococcus multilocularis* introduction and establishment in wildlife via imported beavers. *Veterinary Record* 172: 606. doi: 10.1136/vr.101572.

Lahti, S. (1997) Development, distribution, problems and prospects of Finnish beaver populations (*Castor fiber* L. and *C. canadensis* Kuhl). In Nitsche, K-A. and Pachinger, K. (eds) *Proceedings of the 1st European Beaver Symposium*, Slovak Zoological Society, Bratislava, Slovakia, pp. 61–65.

Lahti, S. and Helminen, M. (1974) The beaver *Castor fiber* (L.) and *Castor canadensis* (Kuhl) in Finland. *Acta Theriologica* 19: 177–189.

Lamsodis, R. and Ulevičius, A. (2012) Geomorphological effects of beaver activities in lowland drainage ditches. *Zeitschrift für Geomorphologie* 56: 435–458.

Lavro, L.S. and Orlov, V.N. (1973) Karyotypes and taxonomy of modern beavers (Castor, Castoridae, Mammalia). *Zoologische Zhurnal* 52: 734–742 (in Russian with English summary).

LeBlanc, F.A., Gallant, D., Vasseur, L. and Leger, L. (2007) Unequal summer use of beaver ponds by river otters: influence of beaver activity, pond size, and vegetation cover. *Canadian Journal of Zoology/Revue Canadienne de Zoologie* 85: 774–782.

Lever, C. (1980) No beavers for Britain. *New Scientist* 80: 812–814.

Lisle, S. (1996) Beaver deceivers. *Wildlife Control Technology* September–October: 42–44.

Lisle, S. (2001) Beaver management at the Penobscot Indian Nation, USA. Using flow devices to protect property and create wetlands. *Proceedings of the 2nd European Beaver Symposium*. Carpathian Heritage Society, Kraków, Poland.

Lisle, S. (2003) The use and potential of flow devices in beaver management. *Lutra* 46: 211–216.

Lokteff, R.L., Roper, B.B. and Wheaton, J.M. (2013) Do beaver dams impede the movement of trout? *Transactions of the American Fisheries Society* 142: 1114–1125.

Longcore, J.R., McAuley, D.G., Pendelton, G.W., Bennatti, C.R., Mingo, T.M. and Stromborg, K.L. (2006) Macroinvertebrate abundance, water chemistry, and wetland characteristics affect use of wetlands by avian species in Maine. *Hydrobiologia* 567: 143–167.

Macarthur, R.A. and Dyck, A.P. (1990) Aquatic thermoregulation of captive and free-ranging beavers (*Castor canadensis*). *Canadian Journal of Zoology* 68: 2409–2416.

Macdonald, D.W., Tattersall, F.H., Brown, E.D. and Balharry, D. (1995) Reintroducing the European beaver to Britain: nostalgic meddling or restoring biodiversity? *Mammal Review* 25: 161–201.

Macdonald, D.W., Maitland, P., Rao, S., Rushton, S., Strachan, R. and Tattersall, F. (1997) Development of a protocol for identifying beaver release sites. Scottish Natural Heritage Research, Survey and Monitoring Report No. 93.

Macmillan, D.C., Duff, E.I. and Elston, D.A. (2001) Modelling the non-market environmental costs and benefits of biodiversity projects using contingent valuation data. *Environmental and Resources Economics* 18: 391–410.

MacTaggart, S.T. and Nelson, T.A. (2003) Composition and demographics of beaver (*Castor canadensis*) colonies in central Illinois. *American Midland Naturalist* 150: 139–150.

Malison, R., Lorang, M.S., Whited, D.C. and Stanford, J.A. (2014) Beavers (*Castor canadensis*) influence habitat for juvenile salmon in a large Alaskan river floodplain. *Freshwater Biology* doi:10.1111/fwb.12343.

Malloch Society (2007) *Managing Aspen Trees for Hammerschmidtia*. http://www.mallochsociety.org.uk/hammerschmidtia-ferruginea/

Manning, A.D., Coles, B.J., Lunn, A.G., Halley, D.J. and Fallon, S.J. (2014) New evidence of late survival of beaver in Britain. *Holocene* 24: 1849–1855.

Mayhew, D.F. (1978) Reinterpretation of the extinct beaver Trogontherium (Mammalia, Rodentia). *Philosophical Transactions of the Royal Society of London. Series B, Biological Sciences* 281: 407–438.

McEwing, R., Senn, H. and Campbell-Palmer, R. (2015) Genetic assessment of wild-living beavers in and around the River Tay catchment, east Scotland. SNH Commissioned Report, No. 682, Battleby.

McKinstry, M.C. and Anderson, S.H. (2002) Survival, fates and success of transplanted beavers, *Castor canadensis*, in Wyoming. *Canadian Field Naturalist* 116: 60–68.

McKinstry, M.C., Caffrey, P. and Anderson, S.H. (2001) The importance of beaver to wetland habitats and waterfowl in Wyoming. *Journal of the American Water Resources Association* 37: 1571–1577.

McNew, L.B., Jr and Woolf, A. (2005) Dispersal and survival of juvenile beavers (*Castor canadensis*) in southern Illinois. *American Midland Naturalist* 154: 217-228.

McRae, G. and Edwards, C.J. (1994) Thermal characteristics of Wisconsin headwater streams occupied by beaver: implications for brook trout habitat. *Transactions of the American Fisheries Society* 123: 641–656.

Metts, B.S., Lanham, J.D. and Russell, K.R. (2001) Evaluation of herpetofaunal communities on upland streams and beaver-impounded streams in the upper Piedmont of South Carolina. *American Midland Naturalist* 145: 54–65.

Mitchell, S.C. and Cunjak, R.A. (2007) Stream flow, salmon and beaver dams: roles in the structuring of stream fish communities within an anadromous salmon-dominated stream. *Journal of Animal Ecology* 76: 1062–1074.

Mörner, T. (1990) *Födseltid hos svenska bävrar (Castor fiber)*. Uppsala: Statens Veterinärmdicinska Anstalt Uppsala (in Swedish with English summary).

Mörner, T., Avenäs, A. and Mattsson, R. (1999) Adiaspiromycosis in a European beaver from Sweden. *Journal of Wildlife Diseases* 35: 367–370.

Morrison, A. (2004) Trial re-introduction of the European beaver to Knapdale: public health monitoring 2001–3. Scottish Natural Heritage Commission Report No. 77.

Müller-Schwarze, D. (2011) *The Beaver: Its Life and Impact*. Ithaca, NY: Cornell University Press.

Müller-Schwarze, D. and Sun, L. (2003) *The Beaver: Natural History of a Wetlands Engineer*. Ithaca, NY: Cornell University Press.

Murphy, M.L., Heifetz, J., Thedinga, J.F., Johnston, S.W. and Koshi, K.V. (1989) Habitat utilisation by juvenile Pacific salmon (*Onchorynchus*) in the glacial Taku River, south-east Alaska. *Canadian Journal of Fisheries and Aquatic Sciences* 46: 1677–1685.

National Species Reintroduction Forum (2014) The Scottish Code for Conservation Translocations and Best Practice Guidelines for Conservation Translocations in Scotland. Scottish Natural Heritage. http://www.snh.gov.uk/protecting-scotlands-nature/reintroducing-native-species/scct/

Naughton-Treves, L., Grossberg, R. and Treves, A. (2003) Paying for tolerance: rural citizens' attitudes toward world depredation and compensation. *Conservation Biology* 17: 1500–1511.

Needham, M.D. and Morzillo, A.T. (2011) Land owner incentives and tolerances for managing beaver impacts in Oregon. Final project report for Oregon Department of Fish and Wildlife and Oregon Watershed Enhancement Board. Oregon State University, Department of Forest Ecosystems and Society, Corvallis, OR.

Nicholson, A.R., Wilkinson, M.E, O'Donnell, G.M. and Quinn, P.F. (2012) Runoff attenuation features: a sustainable flood mitigation strategy in the Belford catchment, UK. *Area* 44: 463–469.

Niinemets, Ü. and Valladares, F. (2006) Tolerance to shade, drought and waterlogging of temperate Northern Hemisphere trees and shrubs. *Ecological Monographs* 76: 521–547.

Niles, J.M., Hartman, K.J. and Keyser, P. (2010) Short-term effects of beaver dam removal on brook trout in an Appalachian headwater stream. *Northeastern Naturalist* 20: 540–551.

Nolet, B.A. (1996) *Management of the Beaver* (Castor fiber)*: Towards Restoration of its Former Distribution and Ecological Function in Europe?* Strasbourg: Council of Europe.

Nolet, B.A. and Baveco, J.M. (1996) Development and viability of a translocated beaver *Castor fiber* population in the Netherlands. *Biological Conservation* 75: 125–137.

Nolet, B.A. and Rosell, F. (1998) Comeback of the beaver *Castor fiber*: an overview of old and new conservation problems. *Biological Conservation* 83: 165–173.

Nolet, B.A., van der Veer, P.J., Evers, E.G.J. and Ottenheim, M.M. (1995) A linear programming model of diet choice of free-living beavers. *Netherlands Journal of Zoology* 45: 317–335.

Nolet, B.A., Broekhuizen, S., Dorrestein, G.M. and Rienks, K.M. (1997) Infectious diseases as main causes of mortality to beavers' (*Castor fiber*) translocation to the Netherlands. *Journal of Zoology* 24: 35–42.

Novak, M. (1987) Beaver. In Novak, M., Baker, J.A., Obbard, M.E. and Malloch, B. (eds) *Wild Furbearer Management and Conservation in North America*. Ontario Ministry of Natural Resources, Toronto, and Ontario Trappers Association, North Bay, pp. 282–312.

Nummi, P. (2001) Alien species in Finland. *The Finnish Environment* 466. Helsinki: Ministry of the Environment, pp. 36–37.

Nummi, P. and Hahtola, A. (2008) The beaver as an ecosystem engineer facilitates teal breeding. *Ecography* 31: 519-524.

Nyssen, J., Pontzeele, J. and Billi, P. (2011) Effect of beaver dams on the hydrology of small mountain streams: example from the Chevral in the Ourthe Orientale basin, Ardennes, Belgium. *Journal of Hydrology* 402: 92–102.

O'Connell, P.E., Ewen, J., O'Donnell, G. and Quinn, P.E. (2007) Is there a link between agricultural land-use management and flooding? *Hydrological Earth System Science* 11: 96–107.

Office Nationale de la Chasse (1997) *Le castor dans le sud-est de la France* [*The Beaver in South-east France*]. Paris: Office Nationale de la Chasse.

Onslow, the Earl of (1939) Why not a National Park in the Highlands? *The Countryman*, January, 496–507.

Pachinger, K. and Hulik, T. (1999) The recent activity of beavers, *Castor fiber*, in the greater Bratislava area. In Busher, P.E. and Dzięciołowski, R.M. (eds) *Beaver Protection, Management, and Utilization in Europe and North America*. New York: Kluwer Academic/Plenum, pp. 53–60.

Paine, R.T. (1995) A conversation on refining the concept of keystone species. *Conservation Biology* 9(4): 962–996.

Parker, A. (2013) Miramichi Salmon Association. Conservation Field Program Report 2013. South Esk, New Brunswick. http://www.miramichisalmon.ca/wp-content/uploads/2012/04/Miramichi-Salmon-Association-Conservation-Report-2013.pdf.

Parker, H. and Rønning, C. (2007) Low potential for restraint of anadromous salmonid reproduction by beaver *Castor fiber* in the Numedalslågen river catchments, Norway. *River Research and Applications* 23: 752–762.

Parker, H. and Rosell, F. (2001) Parturition dates for Eurasian beaver *Castor fiber*: when should spring hunting cease? *Wildlife Biology* 7: 237–241.

Parker, H. and Rosell, F. (2003) Beaver management in Norway: a model for continental Europe? *Lutra* 46: 223–234.

Parker, H. and Rosell, F. (2012) *Beaver Management in Norway – A Review of Recent Literature and Current Problems*. HiT Publication no. 4/2012.

Parker, H. and Rosell, F. (2014) Rapid rebound in colony number of an over-hunted population of Eurasian beaver *Castor fiber*. *Wildlife Biology* 20: 267–269.

Parker, H., Rosell, F. and Holthe, V. (2000) A gross assessment of the suitability of selected Scottish riparian habitats for beaver. *Scottish Forestry* 54: 25–31.

Parker, H., Haugen, A., Kristensen, Ø., Myrum, E., Kolsing, R. and Rosell, F. (2001a) Landscape use and economic value of Eurasian beaver (*Castor fiber*) on a large forest in southeast Norway. In Busher, P. and Gorshkov, Y. (eds) *First Euro-American Beaver Congress, Volga–Kama National Nature Zapovednik, Kazan, Russia, 24–28 August 1999*, pp. 77–95.

Parker, H., Rosell, F., Hermansen, A., Sørløkk, G. and Stærk, M. (2001b) Can beaver *Castor fiber* be selectively harvested by sex and age during spring hunting? In Czech, A. and Schwab, G. (eds) *The European Beaver in a New Millennium. Proceedings of the 2nd European*

Beaver Symposium, 27–30 September 2000, Bialowieza, Poland. Carpathian Heritage Society, Kraków, pp. 164–169.

Parker, H., Rosell F. and Gustavsen, P.Ø. (2002a) Errors associated with moose-hunter counts of occupied beaver *Castor fiber* lodges in Norway. *Fauna Norvegica* 22: 23–31.

Parker, H., Rosell, F., Hermansen, A., Sørløkk, G. and Stærk, M. (2002b) Sex and age composition of spring-hunted Eurasian beaver in Norway. *Journal of Wildlife Management* 66: 1164–1170.

Parker, H., Rosell, F. and Danielsen, J. (2006a) Efficacy of cartridge type and projectile design in the harvest of beaver. *Wildlife Society Bulletin* 34: 127–130.

Parker, H., Rosell, F. and Mysterud, A. (2006b) Harvesting of males delays female breeding in a socially monogamous mammal: the beaver. *Biology Letters* 3: 106–108.

Parker, H., Nummi, P., Hartman, G. and Rosell, F. (2012) Invasive North American beaver *Castor canadensis* in Eurasia: a review of potential consequences and strategy for eradication. *Wildlife Biology* 18: 354–365.

Parker, H., Steifeten, Ø., Uren, G. and Rosell, F. (2013) Use of linear and areal habitat models to establish and distribute beaver *Castor fiber* quotas in Norway. *Fauna Norvegica* 33: 29–34.

Parrot, A., Brooks, W., Harmar, O. and Pygott, K. (2009) Role of rural land use management in flood and coastal risk management. *Journal of Flood Risk Management* 2: 272–284.

Payne, N.F. (1984) Mortality rates of beaver in Newfoundland. *Journal of Wildlife Management* 48: 117–126.

Peck, S.B. (2006) Distribution and biology of the ectoparasitic beaver beetle *Platypsyllus castoris* Ritsema in North America (Coleoptera: Leiodidae: Platypsyllinae). *Insecta Mundi* 20: 85–94.

Pence, L. (1999) Beaver: a tool for riparian management. *Beaver and Common-Sense Conflict Solutions,* Estes Park, Colorado, 7–9 September.

Perryman, H. (2013) Comment on Siemer *et al.* 2013. *Human–Wildlife Interactions* 7: 334.

Piechocki, R. (1977) Ökologische Todesusachenforschung am Elbebiber (*Castor fiber albicus*). *Beiträge Jagd-Wildforsch* 10: 332–341.

Pinto, B., Santos, M.J., Rosell, F. (2009) Habitat selection of the Eurasian beaver (*Castor fiber*) near its carrying capacity: an example from Norway. *Canadian Journal of Zoology* 87: 317–325.

Pitt, M.E. (2008) The Pitt Review – learning lessons from the 2007 floods. http://webarchive.nationalarchives.gov.uk/20100807034701/http://archive.cabinetoffice.gov.uk/pittreview/thepittreview/final_report.html

Pizzi, R. (2014) Minimally invasive surgery techniques. In Miller, R.E. and Fowler, M.E. (eds) *Zoo and Wild Animal Medicine,* vol. 8, pp. 688–698. St Louis, MO: Elsevier Saunders.

Pizzi, R., Campbell-Palmer, R., Cracknell, J., Anderson, T., Brown, D. and Girling, S. (2012a) Minimally invasive surgical screening of Eurasian beavers (*Castor fiber*) for *Echinococcus multilocularis.* British Veterinary Zoological Society, Edinburgh Zoo, 10–11 November 2012.

Pizzi, R., Cracknell, J. and Carter, P. (2012b) Ante-mortem screening of beavers for Echinococcus. *Veterinary Record* 170: 293–294.

Pollock, M.M., Wheaton, J.M., Bouwes, N. and Jordan, C.E. (2011) Working with beaver to restore salmon habitat in the Bridge Creek intensively monitored watershed: design rationale and hypotheses, interim report. NOAA Northwest Fisheries Science Center: Seattle, WA.

Puplett, D. (2008) Beavers and Aspen: looking to the future. *Aspen in Scotland: Biodiversity and Management. Proceedings of a Conference held in Boat of Garten, October 2008,* pp. 27–30.

Ratkus, G.V. (2006) The review of the restoration of migrating fish resources of the republic of Lithuania and means for their implementation. *Symposium on Hydropower, Flood Control and Water Abstraction: Implications for Fish and Fisheries.* Mondsee, Austria.

Reynolds, P. (2000) European beaver and woodland habitats: a review. SNH Review 126, Battleby.

Richard, P.B. (1967) Le déterminisme de la construction des barrages chez le Castor du Rhône. *Terre et la Vie* 21: 339–472.

Roblee, K.J. (1987) The use of T-culvert guard to protect road culverts from plugging damage by beavers. *Proceedings, East Wildlife Damage Control Conference*, vol. 3, pp. 25–33.

Rodgers, J. (1947–48) *English Rivers*. London: Batsford.

Rolauffs, P., Hering, D. and Lohse, S. (2001) Composition, invertebrate community and productivity of a beaver dam in comparison to other stream habitat types. *Hydrobiologia* 459: 201–212.

Rosell, F., Bozsér, O., Collen, P. and Parker, H. (2005) Ecological impact of beavers *Castor fiber* and *Castor canadensis* and their ability to modify ecosystems. *Mammal Review* 35: 248–276.

Rosell, F. and Czech, A. (2000) Responses of foraging Eurasian beavers *Castor fiber* to predator odours. *Wildlife Biology* 6: 13–21.

Rosell, F. and Hovde, B. (1998) Pine marten, *Martes martes*, as a Eurasian Beaver, *Castor fiber*, lodge occupant and possible predator. *Canadian Field-Naturalist* 112: 535–536.

Rosell F. and Hovde, B. (2001) Methods of aquatic and terrestrial netting to capture Eurasian beavers. *Wildlife Society Bulletin* 29: 269–274.

Rosell, F. and Kvinlaug, J.K. (1998) Methods for live-trapping beaver (*Castor* spp.). *Fauna Norv Series A* 19: 1–28.

Rosell, F. and Sun, D. (1999) Use of anal gland secretion to distinguish the two beaver species *Castor canadensis* and *C. fiber*. *Wildlife Biology* 5: 119–123.

Rosell, F. and Thomsen, L.R. (2006) Sexual dimorphism in territorial scent marking by adult Eurasian beavers (Castor fiber). *Journal of Chemical Ecology* 32: 1301.

Rosell, F., Bergan, F., Parker, H. (1998) Scent-marking in the Eurasian beaver (*Castor fiber*) as a means of territory defense. *Journal of Chemical Ecology* 24: 207–219.

Rosell, F., Rosef, O. and Parker, H. (2001) Investigations of waterborne pathogens in Eurasian beavers (*Castor fiber*) from Telemark County, Southeast Norway. *Acta Scandinavica* 42: 479–482.

Rosell, F., Parker, H. and Steifetten, Ø. (2006) Use of dawn and dusk sight observations to determine colony size and family composition in Eurasian beaver *Castor fiber*. *Acta Theriologica* 51: 107–112.

Rosenau, S. (2003) 'Bibermanagementplan' – Entwicklung eines Schutzkonzeptes für den Biber (*Castor fiber* L.) im Bereich der Berliner Havel. *Zwischenbericht* 1–8.

Rudzīte, M. (2005) Assessment of the condition of freshwater pearl mussel *Margaritifera margaritifera* (Linnaeus 1758) populations in Latvia. *Acta Universitatis Latviensis* 691: 121–128.

Rudzīte, M. and Rudzīte, M. (2011) The populations of the freshwater pearl mussel *Margaritifera margaritifera* in Natura 2000 site nature reserve 'Melturu sils'. *Environmental and Experimental Biology* 9: 37–41.

Rudzīte, M. and Znotina, V. (2006) An answer to Campbell. *Acta Universitatis Latviensis* 710: 161–163.

Runyon, M.J., Tyers, D.B., Sowell, B.F. and Gower, C.N. (2014). Aspen restoration using beaver on the northern Yellowstone winter range under reduced ungulate herbivory. *Restoration Ecology* 22: 555–561.

Saveljev, A.P. and Safonov, V.G. (1999) The beaver in Russia and adjoining countries: recent trends in resource changes and management problems. In Busher, P.E. and Dzięciołowski, R.M. (eds) *Beaver Protection, Management, and Utilization in Europe and North America*. New York: Kluwer Academic/Plenum, pp. 17–24.

Saveljev, A.P., Stubbe, M. and Stubbe, A. (2002) Natural movements of tagged beavers in Tyva. *Russian Journal of Ecology* 33: 434–439.

Schlosser, I.J. (1995) Dispersal, boundary processes, and trophic-level interactions in streams adjacent to beaver ponds. *Ecology* 76: 908–925.

Schlosser, I.J. (1998) Fish recruitment, dispersal and trophic interactions in a heterogeneous ionic environment. *Oecologia* 113: 260–268.

Schmidbauer, M. (1996) Bestandsermittlung, Problemanalyse sowie Erarbeitung eines Maßnahmenkonzeptes und dessen Umsetzung zu Bibervorkommen in ausgewählten Oberpfälzer Teichgebieten. Unpublished report, 1–82. http://www.bhl-europe.eu/static/a0fxpnz6/a0fxpnz6_full_ocr.txt

Schulte, B.A. (1998) Scent marking and responses to male castor fluid by beavers. *Journal of Mammalogy* 79: 191–203.

Schulte, B.A. and Müller-Schwarze, D. (1999) Understanding North American beaver behaviour as an aid to management. In Busher, P.E. and Dzięciołowski, R.M. (eds) *Beaver Protection, Management, and Utilization in Europe and North America*. New York: Kluwer Academic/Plenum, pp. 109–128.

Schwab, G. (2014) *Handbuch für den Biberberater*. Loose-leaf collection, available at www.biberhandbuch.de.

Schwab, G. and Schmidbauer, M. (2001) The Bavarian beaver reintroductions. In Czech, A. and Schwab, G. (eds) *The European Beaver in a New Millennium. Proceedings of the 2nd European Beaver Symposium, 27–30 September 2000, Bialowieza, Poland*, Carpathian Heritage Society, Poland, pp. 51–53.

Schwab, G. and Schmidbauer, M. (2003) Beaver (*Castor fiber* L., Castoridae) management in Bavaria. Download from www.biologiezentrum.at

Schwab, G., Dietzen, W. and Lossow, G. (1992) *Biber in Bayern: Entwicklungskonzept zum Schutz des Bibers in Bayern*. München: Bayerisches Landesamt für Umweltschutz, pp. 1–104.

Schwab, G., Dietzen, W. and Lossow, G. (1994) Biber in Bayern, Entwicklung eines Gesamtkonzeptes zum Schutz des Bibers. In Bayerisches Landesamt für Umweltschutz (ed.), *Biber*. München: Schriftenreihe des Bayerischen Landesamtes für Umweltschutz, 128 (Beiträge zum Artenschutz 18), pp. 9–44.

Scottish Environment Protection Agency (2014) WAT-PS-14-01 (The Controlled Activity Regulations) CAR and the management of beaver structures. http://www.sepa.org.uk/regulations/water/engineering/engineering-guidance

Scottish Natural Heritage (1998) *Reintroduction of the European Beaver to Scotland: A Public Consultation*. Battleby: SNH.

Scruton, D., Anderson, T.C. and King, L.W. (1998) Pamehac Brooks: a case study of the restoration of a Newfoundland, Canada, river impacted by flow diversion for pulpwood transportation. *Aquatic Conservation: Marine and Freshwater Ecosystems* 8: 145–157.

Sieber, J. (1999) The Austrian beaver, *Castor fiber*, reintroduction program. In Busher, P.E. and Dzięciołowski, R.M. (eds) *Beaver Protection, Management, and Utilization in Europe and North America*. New York: Kluwer Academic/Plenum, pp. 37–42.

Siemer, W.F., Jonker, S.A., Decker, D.J. and Organ, J.F. (2013) Toward an understanding of beaver management as human and beaver densities increase. *Human–Wildlife Interactions* 7: 114–131.

Sigourney, D.B., Letcher, B.H. and Cunjak, R.A. (2006) Influence of beaver activity on summer growth and condition of age-2 Atlantic salmon parr. *Transactions of the American Fisheries Society* 135: 1068–1075.

Simon, L.J. (2006) Solving beaver flooding problems through the use of water flow control devices. In: *Proceedings of the 23rd Vertebrate Pest Conference, University of California, Davis*. pp 174–180.

Simpson, V. and Hartley, M. (2011) Echinococcus risk from imported beavers. *Veterinary Record* 169: 689.

Sjöberg, G. (1998) Ecosystem engineering in forest streams, invertebrate fauna in beaver ponds (abstract). *European–American Mammal Congress, 19–24 July 1998, Santiago de Compostela, Spain*. Santiago de Compostela, Universidad de Spain, p. 158.

Smith, D.H. and Peterson, R.O. (1988) The effect of regulated lake levels on beaver in Voyageurs National Park, Minnesota. US Department of Interior, National Park Service, Research/Resources Management Report MWR-11. Midwest Regional Office, Omaha, NE.

SNH (2015) Beavers in Scotland: a report to the Scottish Government. Scottish Natural Heritage. www.snh.gov.uk/beavers-in-Scotland.

Snodgrass, J.W. and Meffe, G.K. (1999) Habitat use and temporal dynamics of blackwater stream fishes in and adjacent to beaver ponds. *Copeia* 628–639.

South, A., Rushton, S. and Macdonald, D. (2000) Simulating the proposed reintroduction of the European beaver (*Castor fiber*) to Scotland. *Biological Conservation* 93: 103–116.

Stich, W. (2005) Beavers in Vienna. *Water: A Worldwide Panorama.* http://worldwidepanorama. org/wwp/rss/go/n 1194.

Stringer, A.S. and Gaywood, M. In press. The impacts of beavers on biodiversity and the ecological basis for reintroduction to Scotland. *Mammal Review.*

Stringer, A.S., Blake, D. and Gaywood, M.J. (2015) A geospatial analysis of potential Eurasian beaver (*Castor fiber*) colonisation following reintroduction to Scotland. Scottish Natural Heritage Commissioned Report No. 875.

Suzuki, N. and McComb, B.C. (2004) Associations of small mammals and amphibians with beaver-occupied streams in the Oregon coast range. *Northwest Science* 78: 286–293.

Swanston, D.N. (1991) Natural processes. In Meehan, W.R. (ed.) *Influences of Forest and Rangeland Management on Salmonid Fishes and Their Habitats.* American Fisheries Society Special Publication 19, Bethesda, MD, pp. 139–179.

Tambets, M., Jarvekulg, R., Veeroja, R., Tambets, J. and Saat, T. (2005) Amplification of negative impact of beaver dams on fish habitats of rivers in extreme climatic condition. *Journal of Fish Biology* 67: 275–276.

Taylor, B.R., Macinnis, C. and Floyd, T.A. (2010) Influence of rainfall and beaver dams on upstream movement of spawning Atlantic salmon in a restored brook in Nova Scotia, Canada. *River Restoration Application* 26: 183–193.

Tayside Beaver Study Group (2015) Final report. http://www.snh.gov.uk/ protecting-scotlands-nature/beavers/tayside-beaver-study-group/

Thomsen, L.R., Campbell, R.D. and Rosell, F. (2007) Tool-use in a display behaviour by Eurasian beavers (*Castor fiber*). *Animal Cognition* 10: 477.

Thorstad, E.B., Økland, F., Aarestrup, K. and Heggberget, T.G. (2008) Factors affecting the within-river spawning migration of Atlantic salmon, with emphasis on human impacts. *Reviews in Fish Biology and Fisheries* 17: 345–371.

Tinnesand, H.V., Jojola, S., Zedrosser, A. and Rosell, F. (2013) The smell of desperadoes? Beavers distinguish between dominant and subordinate intruders. *Behavioral Ecology and Sociobiology* 67: 895–904.

Törnblom, J., Angelstam, P., Hartman, G., Henrikson, L. and Sjöberg, G. (2011) Toward a research agenda for water policy implementation; knowledge about beaver (*Castor fiber*) as a tool for water management with a catchment perspective. *Baltic Forestry* 17: 154–161.

Tyurnin, B.N. (1984) Factors determining numbers of the river beavers (*Castor fiber*) in the European North. *Soviet Journal of Ecology* 14: 337–344.

Ulevičius, A. and Janulaitis, M. (2007) Abundance and species diversity of small mammals on beaver lodges. *Ekologija* 53: 38–43.

van den Brink, F.H. (1967) *A Field Guide to the Mammals of Britain and Europe.* London: Collins.

Vorel, A., Šíma, J. *et al.* (2013) *Program péče o populaci bobra evropského v České republice.* Praha: AOPK ČR, p. 99.

Vuitton, D.A., Zhou, H., Bresson-Hadni, S., Wang, Q., Piarroux, M., Raoul, F. and Giraudoux, P. (2003) Epidemiology of alveolar echinococcosis with particular reference to China and Europe. *Parasitology* 127: 87–107.

Wagner, T.B., Kimberly, K. and Dale, L. (2000) Evaluation of Hot Sauce® as a repellent for forest mammals. *Wildlife Society Bulletin* 28: 76-83.

Walsh, C.L. and Kilsby, C.G. (2007) Implications of climate change on flow regime affecting Atlantic salmon. *Hydrology and Earth System Science* 11: 1125–1141.

Webb, A., French, D.D. and Flitsch, A.C.C. (1997) Identification and assessment of possible beaver sites in Scotland. Scottish Natural Heritage Commissioned Report, No. 94, Edinburgh.

Westbrook, C.J., Cooper, D.J. and Baker, B.W. (2006) Beaver dams and overbank floods influence groundwater–surface water interactions of a Rocky Mountain riparian area. *Water Resources Research* 42: doi:10.1029/2005WR004560.

Wilkinson, M.E., Quinn, P.F. and Welton, P. (2010) Runoff management during the September 2008 flood in the Belford catchment, Northumberland. *Journal of Flood Risk Management* 3: 285–295.

Willby, N.J., Casas Mulet, R. and Perfect, C. (2011) The Scottish Beaver Trial: monitoring and further baseline survey of the aquatic and semi-aquatic macrophytes of the lochs, 2009. Commissioned Report 455, Scottish Natural Heritage, Perth.

Willby, N., Perfect, C. and Law, A. (2014) The Scottish Beaver Trial: monitoring of aquatic vegetation and associated features of the Knapdale lochs 2008–2013, final report. Scottish Natural Heritage Commissioned Report No. 688.

Wilsson, L. (1971) Observations and experiments on the ethology of the European Beaver (*Castor fiber* L.). *Viltrevy* 8: 115–166.

Wisconsin Department of Natural Resources (WI DNR) (2005) Beaver damage control: Guidelines for people with beaver damage problems. Publ WM-007-05 REV. Madison, WI.

Wobeser, G., Campbell, G.D., Dallaire, A. and McBurney, S. (2009) Tularaemia, plague, yersiniosis and Tyzzer's disease in wild rodents and lagomorphs in Canada: a review. *Canadian Veterinary Journal* 50: 1251–1256.

Wohl, E. (2013) Landscape-scale carbon storage associated with beaver dams. *Geophysical Research Letters* 40: 1–6.

Wood, D.M. (1965) Studies on the beetles *Leptinillus validus* (Horn) and *Platypsyllus castoris* Ritsema (Coleoptera: Leptinidae) from beaver. *Proceedings of the Entomological Society of Ontario* 95: 33–63.

Wood, G.W., Woodward, L.A. and Yarrow, G.K. (1994) The Clemson beaver pond leveller. AFW Leaflet 1, Clemson Coop. Ext. Service, Clemson, SC.

Wright, J.P., Jones, C.G. and Flecker, A.S. (2002) An ecosystem engineer, the beaver, increases species richness at the landscape scale. *Oecologia* 132: 96–101.

Żurowski, W. and Kasperczyk, B. (1986) Characteristics of a European beaver population in the Suwałki lakeland. *Acta Theriologica* 31: 311–332.

Żurowski, W. and Kasperczyk, B. (1988) Effects of reintroduction of European beaver in the lowlands of the Vistula basin. *Acta Theriologica* 33: 338–338.

Glossary

Aggradation: to fill and raise the level of a water body by deposition of sediment.

Agri-environment schemes: voluntary land-management agreements that provide funding for farmers and land-managers who manage their land in an environmentally sensitive way above and beyond standard regulations.

Anadromous salmonids: species that are born in fresh water, migrate to the sea where the majority of their life is spent, before returning to fresh water to spawn and die.

Anal gland secretions: from a pair of internal sebaceous glands, anal papillae can be exposed through the cloaca to secrete a liquid that can be used in territorial defence and for identification of family members. Colour and viscosity can be used to sex beavers.

Anthropogenic: the effect or object resulting from human activity.

Beaver control/management: mitigation of beaver impact, including a range of non-lethal and lethal control methods.

Biomass: organic matter derived from living, or recently living organisms.

Cadaver: corpse or deceased body.

Carbon sequestration: long-term storage of carbon dioxide or other forms of carbon to mitigate or defer global warming.

Castoreum: yellowish secretion of the castor sacs. Beavers use castoreum in combination with urine to scent-mark territory.

Castration: removal of the testicles, to prevent a male from reproducing.

Chemical communication: a form of communication between animals based on chemicals or pheromones being emitted or deposited for the purpose of signalling various messages to others of the same species, such as readiness for mating, boundaries of territories and identity of individuals in an area. These signals can also be picked up by non-target species: for example, prey animals can recognise the presence of a predator and may then leave an area; and predators can track prey.

Colonisation/recolonization: process by which a species spreads to new areas or repopulates a previously occupied area it became expatriated from.

Cultural landscapes: cultural properties of an area that represent both natural and anthropogenic factors.

Diptera: true flies.

Domestic legislation: law relating to a country's internal affairs.

Dynamic environment: constantly changing environment.

Dynamic source-sink pattern: a theoretical model used to describe how variation in habitat quality may affect the population growth or decline of organisms.

Ecosystem services: processes by which the environment produces resources that are utilised by humans.

Euthanasia: the act of putting an animal to death or allowing it to die by withholding extreme medical measures.

Eutrophication: the enrichment of an environment as a result of excessive nutrient input, commonly in low nutrient water systems.

Extant: still in existence; surviving.

Fluke: internal, parasitic flatworm.

Geographic expansion: the dispersal of a species into a surrounding landscape.

Habitat heterogeneity: habitats that contain a greater number and diversity of niches thereby supporting more biodiversity.

Habituation: a form of learning in which an individual learns to stop responding to a stimulus which is no longer biologically relevant.

Herbivorous: species adapted to a plant diet.

Host-specific: a parasite that is capable of living solely on or in one species of host.

Hypothermia: the condition of having an abnormally (typically dangerously) low body temperature.

Impoundment: to collect water behind a dam or in a reservoir.

Keystone species: a species that has a disproportionately large effect on its environment relative to its abundance.

Kits: the young of beavers.

Lactation: the secretion of milk in female mammals.

Limiting factor: factors preventing a population from growing.

Migratory fish: fish that migrate from one location to another for either feeding or breeding purposes.

Natal: place of birth.

Native range: refers to where a species historically originated and lived.

Nematode: roundworms, many of which are parasitic.

Non-native: a species living outside its native distributional range.

Opportunistic predators: species that feed on a range of prey when opportunity presents itself.

Ovariectomy: removal of one or both ovaries.

Ovariohysterectomy: removal of the uterus and ovaries

Parr: a young salmonid species, between the stages of fry and smolt.

Pathogen/pathogenic: causing or capable of causing disease.

Pelage: the fur, hair, wool, etc., of a mammal.

Percid: the collective term for perch fish species.

Populations: all the animals or plants of a particular species present in an area.

Preventative measures: action to avoid, minimise or reduce the extent of a problem(s).

Refugia: relict population.

Regional extirpation: the extinction of a species within an area.

Riparian: the area of land adjacent to a water body.

Runoff attenuation features: features which aid in the slowing of water flow along a catchment.

Salmonids: a collective term for the family of ray-finned fish, which includes salmon, trout, chars, freshwater whitefishes and graylings.

Self-regulating population: a population whose growth is naturally limited as a result of various factors such as lack of suitable habitat and food, deaths as a result of territorial fighting, etc.

Semi-emergent: plant species which have roots under water and leaves and flowers above the surface.

Soft engineering: the use of ecological principles and practices to reduce erosion and achieve the stabilisation and safety of shorelines and the area surrounding rivers, while enhancing habitat and improving aesthetics.

Statutory nature conservation organisations: nature conservation organisations which are run by the government.

Sterilisation: eliminating reproduction by physically or chemically altering the function of the sex organs.

Suboptimal habitats: less desirable environments.

Territory: an area that an animal considers as its own and that it defends against intruders of the same species.

Translocation: movement, or the act of moving an animal from one location to another.

Trematode: a parasitic flatworm.

Vasectomy: a surgical operation in which the vas deferens from each testis is cut and tied to prevent transfer of sperm during ejaculation.

Zoonoses/zoonotic: a disease that can be transmitted from vertebrate animals to humans.

Index

Page numbers in *italic* indicate figures or tables and in **bold** indicate glossary terms.